IN EXTREME DANGER

First edition published in 2006 by Awa Press, 16 Walter Street,
Wellington, New Zealand

National Library of New Zealand Cataloguing-in-Publication Data
Mackley, Geoff, 1963-
In extreme danger : Chasing and filming natural disasters and catastrophic weather across the globe / by Geoff Mackley ; with John McCrystal.
Includes index.
ISBN-13: 978-0-9582509-8-6
ISBN-10: 0-9582509-8-7
1. Mackley, Geoff, 1963- 2. Photographers—New Zealand—
Biography. I. McCrystal, John. II. Title.
770.92—dc 22

Awa Press acknowledges with thanks the United States National Oceanic and Atmospheric Agency (NOAA) for the satellite imagery reproduced in this book.

This book is typeset in Palatino

Designed and typeset by Jason Anscomb
Printed by Midas Printing International Ltd, China

www.awapress.com

IN EXTREME DANGER

CHASING AND FILMING NATURAL DISASTERS AND CATASTROPHIC WEATHER ACROSS THE GLOBE

Geoff Mackley

With JOHN McCRYSTAL

AWA PRESS

Geoff Mackley

Geoff Mackley is one of the world's leading photographers of extreme natural disasters and weather events. His amazing close-up film footage of hurricanes, cyclones, typhoons, earthquakes, tsunamis, volcanic eruptions, tornadoes and big chills leads news broadcasts around the world, he is the star of Discovery Channel series *Dangerman*, and his daring exploits have featured in numerous television documentaries and media stories.

After starting his career as a news cameraman, Mackley leapt to world prominence in 1995, when a huge volcano on New Zealand's central plateau suddenly exploded in a series of violent eruptions. As everyone else fled down the mountain, Mackley grabbed his camera and climbing gear and headed for the crater, evading a police cordon and scaling the erupting volcano in pitch-darkness and freezing cold.

The footage he took was powerful and mesmerising – pure television magic. It sold to networks across the globe. Geoff Mackley was on his way to international fame.

Today, whether there's a Category 5 hurricane in the United States, a supertyphoon in the Philippines or a severe tropical cyclone in Australia, Geoff Mackley will be there, lashing himself to a telephone pole and clapping a viewfinder to his eye.

Whenever a volcano blows its top – in Hawai'i, Indonesia, Italy, Costa Rica, Ethiopia, the South Pacific's fabled 'rim of fire', or anywhere else – the figure silhouetted against the fountain of lava will be his. If a Magnitude 9 earthquake or a devastating tsunami hits a region, no matter how remote, he will be one of the first outsiders to arrive on the scene.

Geoff Mackley's fearlessness in the face of nature's full fury, and his astonishing bravery in putting himself in the hot seat, has taken him to the very top of the ranks of international extreme cinematographers. Here, he shares with us the thrill and the terror of a life spent getting himself and his camera in harm's way. What makes a person race headlong into danger? This book will be a revelation.

contents

Life is either a daring adventure

or

nothing

HELEN KELLER

Teenage Fire-chaser to Television Cameraman

I was born in Christchurch, in the South Island of New Zealand. New Zealanders pride themselves on being an outdoor people and this is certainly true of South Islanders Christchurch lies at the margin of the alluvial plains set down by the great braided rivers which flow from the Southern Alps to the sea. It is within reach of just about every imaginable outdoor activity. The alps themselves are an hour's drive away; you can be sipping a cappuccino outside a café in Cathedral Square one minute and tramping through the bush of the foothills 45 minutes later. Keep driving for another half-hour and you'll reach the alpine village of Arthur's Pass. Yank on your boots, and even in summer you can be above the snowline after an hour and a half's walking. Another couple of hours if you're fit and you can be at the summit of a 2275-metre peak, eyeballing Mount Cook, New Zealand's highest mountain.

If you carry on driving for two more hours, you will be in a different landscape again: West Coast rain forests, threaded through with gold-bearing, ice-fed rivers. Turn north, and in three hours you will be in limestone country, pocked with sink-holes and honeycombed with caves. Turn south, and in three hours you can walk on the terminal face of a glacier. And that's only the South Island. A few hours' drive, followed by a ferry ride or an even shorter plane ride, will bring you within striking distance of the North Island's volcanoes, geysers and hot springs.

Filming the big chill, Southern Alps, New Zealand, 1993.

Because the South Island is a high piece of land smack in the middle of the Roaring Forties, it cops the weather. The West Coast gets the brunt: it is subject to staggering rainfall at the best of times. At the worst, when a larger-than-usual system romps in from the Tasman Sea, the river catchments are deluged, and the rivers overrun their banks.

Most of the rain stays on the western side of the great divide, and the only sign we get of it in Christchurch is a dramatic ceiling of orographic cloud, the famous Canterbury nor'west arch. From time to time, though, a storm comes from the south or east or gets over the alps, and our local rivers – the Waimakariri, the Rangitata and the Rakaia – are transformed from a tangle of placid rivulets to roaring torrents of angry brown water, rolling boulders and hurling logs. On one famous occasion even the Heathcote, the tame little creek that snakes through my suburb – across the road, in fact, from the house in which I grew up – burst its banks and overran the neighbourhood. And there's always at least one day a year when you wake up and look out your window to find the city transformed by a cloak of fresh snow.

I was lucky enough to be born into an adventure-minded family, the oldest of three boys. My father, Doug, has always been a keen outdoorsman. Until recently he worked as a customs agent. For most people, this conjures up images of men in uniforms at airports, stamping passports, asking questions and rifling through your luggage, with the threat of the rubber glove hanging in the air. In reality that has nothing to do with the work of customs agents, who complete paperwork requirements for importers and exporters. It's a sedentary desk job, which is perhaps why Dad has always taken such an interest in the natural world. And from my earliest days, I noticed that he was fascinated with the sky.

When there were storms and most of our neighbours were inside huddling in front of the fire, Dad would have my brothers and me out on the lawn, watching the lightning and calculating, by the time it took for the thunder to follow the flash, how far away the centre of the storm was. Whenever there was a flood, we would be standing at a safe distance watching the wild water. Whenever a big dump of snow blanketed the foothills of the alps, we would be up there, chucking snowballs and marvelling at a landscape transformed overnight.

At every opportunity my father and mother, Marie, took us on day-trips and camping expeditions. As we got older and stronger, these trips became more adventurous. We had a Morris Oxford van, a sturdy old workhorse with plenty of room for kids and camping gear and, man, did we cover some ground in that old thing. When it finally gave up the ghost, we moved up to a Bedford J4, the vehicle of choice for touring rock 'n' roll bands in the 1970s. On one memorable trip to south Westland, glacier country, we walked up the track towards the terminal face of the Franz Josef glacier to find a fairly substantial berg had broken off. It had also been a week of heavy rain, and the river bed was littered with refrigerator-sized blocks of dirty ice. We could touch rainwater which had been frozen in the nevée high up there in the Southern Alps before human beings came to New Zealand.

ABOVE
An early adventure in a Ja Class steam locomotive in Christchurch railway yard, June 1969.

RIGHT
With my mother and brothers Steven (middle) and Richard (right) at Franz Josef Glacier, December 1971.

When he thought we were old enough, Dad took each of us in turn on a tramping trip. Mine was to the Heaphy Track. This classic walk from Westland to Golden Bay is one of New Zealand's most famous. It was four days' solid walking and hard work for a 12-year-old, but my father was a great companion, observant and knowledgeable, and certainly one of the reasons I carried a fascination with the natural world into my teenage years. The following year, my first at high school, a couple of schoolmates and I caught a train to Arthur's Pass and climbed 2408-metre-high Mount Murchison. I can't remember if we omitted to tell our parents what we proposed to do, or whether they just decided it was futile trying to stop us, but it was a staunch climb.

On my first major tramping trip, the four-day Heaphy Track in 1975. The Heaphy, one of New Zealand's famous Great Walks, runs through spectacular wilderness to the wild seas of the West Coast.

My friends and I had an equally hair-raising time on a caving trip in Westland's Punakaiki Valley. We were in a cave system when the water came up, as it can do very quickly in the steep, high-rainfall West Coast catchments. Trapped, we were lucky enough to find a chamber above the reach of the flood, and were stuck there for a few hours before we could escape.

The other interest my parents instilled in me was photography, and I soon began to pursue this in my own right. I inherited an ancient Box Brownie camera and took it everywhere. I was fascinated with trains, and spent a lot of my spare time cycling off to the railway and photographing engines – a good introduction to the basics of photography. I was camera-mad, as were my two best friends. We would spend lunch-times in the school library, poring over photography magazines and the camera ads in *Time* and *Newsweek*.

Although my schooling was unremarkable, one primary school teacher stands out in my memory. Jim Coxon had clearly cottoned on to my fascination with natural phenomena, because whenever the senior pupils were having one of their regular movie screenings – often documentaries on volcanoes, hurricanes or earthquakes – he'd invite me to join them. I remember as clearly as if it were yesterday watching the famous footage shot by the pioneering French volcanologists and cinematographers Maurice and Katia Krafft of a volcano devastating the town of Vestmannaeyjar, on the small island of Heimaey in Iceland. I was spellbound.

Later, at high school, I tended to do better in practical subjects such as engineering, woodwork and science than in more abstract ones. I could never see the point of half of what we were taught. If there was one, it was never made plain. You could keep maths, for example. I never understood the theorem of Pythagoras. I still don't. I have to say that particular gap in my knowledge hasn't really held me back.

Most New Zealanders remember their school-days for team sports – particularly rugby – but I preferred individual sports, especially long-distance and cross-country running. I suppose I was influenced here too by my father, a keen runner. With his encouragement, I joined harriers and athletics clubs. I enjoyed running, not for the competitive aspect but for the pitched battle you fought against yourself, finding your limits and seeing how far you could push them.

By my fourth year in high school, even the joys of engineering and woodwork had worn off; I was far more interested in what I could get up to in the weekends. I had already found myself part-time jobs to fund tramping, caving and climbing trips, and to save doggedly for that flash SLR camera I dreamed of owning. Like most kids, I'd started with a paper round. Later on, I had got a horrible job in a butchery hosing down the bloody trays and floor at the end of the day. The reward came the day I found I had finally saved enough to buy the camera, a 35mm Nikon EM with a standard 50mm lens, just about the best you could get in those days and certainly good enough for press photographers and professionals. Not that I imagined it would become the tool of any kind of trade for me: photography was just a hobby.

At the end of the year I decided to leave school, and took the first suitable job I could find, as an apprentice baker at the now long-gone Boon's Bakery. After completing a three-month trade certificate, I started work in earnest. The hours were rugged: I started at eleven at night and went through to mid-morning. I used to go to the bakery on my push-bike, pedalling through the empty streets in the dark. Some days, when the weather was stormy or snowy, that ride to work was the reward for the long night hours. At those times, I loved it.

The first step on my career path came early one morning in April 1980, shortly after I had begun the bakery job. A series of loud bangs woke our neighbourhood, and looking north you could see a column of black smoke rising. It was

I had just turned 17 when I sold my first photographs. A brickworks not far from my home caught fire, and I was first on the spot. Later, factory staff called wanting to buy copies of my photos, and I sold them for a few dollars each.

lit orange and red from beneath. Fire! Without stopping to change out of my pyjamas, I grabbed my camera and a jacket, jumped on my bike and headed for the action. Murphy's Brickworks was going up in flames. A brickworks doesn't sound very flammable but take my word for it, it was burning. Flames were leaping from every orifice, corrugated iron was glowing cherry-red, and every now and then explosions or collapsing timbers would send showers of sparks soaring into the night sky.

When I arrived there was a fair-sized crowd, but not the Fire Service and definitely not the media. I began taking pictures, and when the fire-fighters eventually turned up I photographed them going into action. I used up all my film. It was a buzz capturing the action, and an even greater one getting my photographs back from the processor and comparing them to the pictures the local paper carried a day later. Some sleepy *Press* photographer had gone around during gentleman's hours, long after the fire was over. His shots showed a couple of fire-fighters playing hoses on a heap of twisted iron, and a few wisps of steam rising. My photographs were much better, showing the flames at their height in full, glorious colour, the grim struggle of the men tackling them, the futility of their efforts against a blaze. You could almost feel the heat looking at them.

It didn't occur to me to try to sell the pictures, and I wouldn't have had a clue where to begin. However, someone on the staff of the brickworks had seen me photographing, and I began getting calls from staff members wanting copies. Making a few bucks was quite gratifying, but it still didn't dawn on me that this might ever amount to more than a hobby. It did, however, occur to me that it was a hobby worth pursuing. I was hooked on the rush I'd got from being in the thick of things with my camera. I started to think about how I could get to the site of other emergencies in time to get the action while it was hot. I figured the Fire Service must use radio to communicate, so it would be possible to listen in. Once, I managed to pick up the service on a broadcast radio, but in the absence of a squelch control there was a terrible din of static whenever no one was on-air.

I went looking for a better solution. At a trade fair I met a man who was demonstrating Tait radios – the kind the Fire Service used – and told him what I wanted to do. For a hundred bucks, he told me, he could make up one of these radios. The price was steep for someone on the kind of money I was earning, but it was state-of-the-art equipment, and I bit the bullet and shelled out. After that, I could get the fire channel loud and clear. A whole new world opened up. I found it addictive to listen to the dramas unfolding all round the city and the country when things were dead quiet at home.

I didn't have long to wait before I got another chance to be first at a fire. One afternoon I was reading in my room with the radio on, tuned as usual to the Fire Service frequency, when I heard a call go over. I didn't pay much attention at first, but then the address set bells ringing: the fire was just around the corner, closer to me than to the fire station. My heart started thumping. I grabbed my camera, dashed out the door and jumped on my bike. All the way there I could see the column of black smoke. At the end of the long driveway, a small crowd of onlookers, mostly neighbours, was watching the house burn. It was a typical suburban house, built of timber like most New Zealand houses, and it was well ablaze. Flames were pouring from most of the windows. Every now and then another pane of glass would pop and another jet of flame roar out. It was an electrifying sight. I dropped my bike and began taking photographs. After a couple of minutes, I noticed a flurry of activity at the back door. One of the neighbours had run out his garden hose and was playing it on the porch. Others were trying to advance on the door. It was then I learned there were people trapped inside.

I put down my camera. Perhaps something could be done. I tried getting to the door, but it was hopeless: I'd get to within about ten feet and it would feel as though my jacket were melting on to me. The man with the hose tried soaking me as I advanced, but that was worse. The water turned to steam, and steam transmits heat to your skin better than air. Even if the people inside stood any chance of surviving, there was no chance of getting close to the house, let alone getting inside. I gave up, picked up my camera and resumed taking pictures.

When the firemen arrived, I filmed them getting to grips with the inferno. It's amazing how you can go unnoticed in such situations: people are so focussed on the mayhem, they don't even seem to see you. It was as though I were invisible. It was only after the firemen had got the blaze under control – breathtakingly fast – and established there really was someone in the house, that everyone, including me, was moved behind a cordon.

It was a funny feeling knowing someone had burned to death only a few metres away from where I was standing, with nothing to be done. I thought about it in the days afterward, but I didn't dwell on it. Even at that age, I suppose, my attitude was that people die. This might have been a particularly horrible way to die and I might have been nearby, but people are dying horribly all around the world, all the time, and I can't do anything for them either. Feeling bad isn't going to help any of them.

This time, I contacted *The Press*. After a bit of confusion and a lot of waiting, I was connected to the illustrations editor. I told him what I thought I had on film. He was keen to have a look. I went in and handed over the little canister. He promised to give me a set of prints and to pay for any pictures that were used. I was handed a replacement roll of film. In the end, the newspaper ran one of the pictures. I got $10. It wasn't a fortune, but compared to the money I was getting at the bakery, and considering it was something I had done for the thrill of it rather than to turn a dollar, it wasn't bad. The thrill was compounded by having my first published piece of photojournalism.

I continued to monitor the fire service channel, and worked out a zone within which it was worth my while trying to get to the scene of emergencies. Arrive there after the fire-fighters and there was usually nothing to see: they were amazingly efficient. So you had to get there first. It was clear I needed greater mobility.

I had got my driver's licence as soon as I turned 15, but on my wages buying a car was out of the question. I managed to scratch together enough to buy a motorbike, a Honda XR250 – a cross-bike, capable of getting me along on the road but also handling light off-road work. This dramatically expanded my action zone. Even in the early days, I had a good record of getting to the scene first. I got to know some of the fire-fighters by sight – and they got to know me. One day I heard one on his radio describing what they'd found when they'd arrived at the scene of a fire.

A motorbike changed my scope of operations, allowing me to get to the scene of emergencies faster – often ahead of fire-fighters, police and ambulances. Income from the photos allowed me to purchase better radio equipment, which I mounted on the back of the bike.

'A few rubberneckers,' he said. 'Oh yeah, and Flash was here.'

It turned out I had got this nickname because the bursts of my flashgun were often the first thing they saw when they pulled up at an emergency. At night they could see the flash from miles away. Some of the senior firemen weren't so amused by my punctuality. A chief told me it was impossible for me to be so consistently beating his highly trained, professional fire-fighters to the scene: I had to be setting the fires.

By now I'd also worked out that if I could monitor the fire service communications I could also monitor the other emergency channels. Soon I was listening in to police, ambulance and even the coastguard. As well as keeping up with man-made emergencies, I would learn about natural disasters long before they hit the news. I went out whenever I thought I could reach a scene in time to get worthwhile pictures. Fires and car accidents were the most common. Plenty of the car crashes had fatalities. These can shake you up, but the viewfinder puts everything at a remove. It's somehow easier when you're looking at the scene in terms of composition and light: it becomes a job you're doing, not something you're there to experience emotionally.

On one occasion, I heard that a car had gone into a swollen river near our house. When I reached the scene the car, with only its roof showing, was floating down the stream. There was clearly someone inside: you could hear him banging on the roof. A crowd of people were standing on the bridge watching. It was a great opportunity for rescue pictures, but something was missing – the rescue! I took off my outer clothes and jumped in, just as the first fire engine arrived. As I reached the car two or three firemen joined me, and we all yanked on the door. It was a good feeling to help in the rescue – we got the guy out alive – but all the while I was wishing I were on the bank taking photos; they were some of the best pictures I never got. It turned out the driver had had a fight with his girlfriend and driven into the river deliberately to try and end it all. He must have changed his mind pretty quickly, because when I got to him he was sure as hell keen to go on living.

By now I was getting photos in the local newspapers regularly, sometimes once a day. With this nice supplement to my income I was able to buy more and better radio equipment, including a set I could mount on my motorbike so I could be still in touch with emergency channels when attending an emergency, or just roaring around the countryside. Soon I was able to afford a new and better motorbike too, a 500cc. The problem with the 500cc motorbike was that it was so fast I got disqualified from driving for overtaking a fire engine – on the way to a fire. Later I got a Series IIA Land Rover. Now we were talking. You can go just about anywhere in a Land Rover and I did. If the weather looked as though it were going to cut up rough on the West Coast, about 250 kilometres away, I would cruise over in time to see the rain start falling. If there were a heavy snowfall in South Canterbury, the Rover took me there in relative comfort and safety.

I was also keeping abreast of new technology and in 1990, as soon as I could afford it, I bought a video camera. The technology has changed so much and so quickly since then that most people have probably forgotten what the first affordable video cameras were like. We're now so used to palm-sized digital camcorders that even their light, compact, 8mm magnetic-tape predecessors are a vague memory. Go back even further and you're in real dinosaur country. My first video camera was the size of a small suitcase. I would feed this contraption a full-sized VHS tape, heave it on my shoulder, clap the rubber viewfinder cup to my eye, and squeeze the trigger on the pistol grip. The quality was rubbish – I could have painted a better picture – but at the time the process seemed little short of a miracle.

By now I had left the bakery and gone to work driving trucks. I had been mad on trucks since I was knee-high but the job was short-lived. Soon after I started, the company I was working for was taken over by an Australian firm, and six months later it was a victim of the rationalisations that swept through the trucking industry. It folded and I was out of a job. I bought an old wreck of a van and got work as an owner-driver for a courier company called Ivory Freight. I can't say I was the best employee it ever had. I was constantly late with deliveries because I'd fitted the van with a scanner and used to monitor the emergency frequencies while I was out and about. Usually I had a good excuse: the radio let me know where there were accidents, fires, floods, or anything else that might explain why I had been caught up. In fact, though, I was far less often caught up than racing to get there, or actually filming.

The breakthrough came one night when, armed with my new video camera, I got to the scene of a fire which was destroying a former nurses' home at a local hospital and filmed the whole event. As with my first still photographs, I had no ambition to do anything with this footage. I had shot it for my own interest and to show my friends and family. However, it so happened that one of my friends who was doing a broadcasting course popped into my home that morning on his way to a lecture.

I grabbed him. 'Have a look at this.'

He watched the footage – the building going up in flames from end to end, the explosions and showers of sparks, the fire-fighters racing about – and turned to me. 'Hell,' he said. 'You could get that on the news.'

'Rubbish,' I replied.

'No, you *will* get that on the news. Trust me.'

Still sceptical, I rang the local office of Television New Zealand. The phone rang and rang and no one answered. It looked as though I was right: it was all pie in the sky. Then I tried the private channel, TV3. Someone answered and I was put through to Mark Jennings, the chief of the local news bureau. I told him I had footage of the hospital fire.

'Do you have flames?' he asked.

'Of course I have flames,' I said. 'The building was burning from end to end.'

'We'd be interested in having a look,' he said.

I raced into TV3 with the tape and Jennings watched it.

'We'll take it,' he said. 'We'll give you three hundred bucks.'

Wow, I thought, you could almost make a living out of this.

It was the first video footage I ever sold, and it was the lead item on the news that night. Afterwards, Jennings congratulated me on its quality, and encouraged me to bring in more.

I rapidly realised that television news was usually short of pictures of emergencies, simply because no one was filming these events *as they happened*. And TV news is driven by images. If a channel doesn't have pictures, a story has to be pretty massive before it'll be reported. Conversely, if an event is small-scale but the newsroom has great pictures, it will get covered. With wars, famines, natural disasters, political revolutions and elections going on around the world every day, house fires shouldn't rate. However, you would be surprised how much prime time gets devoted to them if there is good footage.

I calculated that it would take only one or two incidents a week to make emergency-chasing a viable career option. It happened that around this time Christchurch was going through a strange phase. A big trial involving members of a scary skinhead gang was about to take place, and other members of the gang were waging a terror campaign to try and intimidate witnesses and jurors. Over the next month there was a constant stream of fires and car explosions. In one incident, a pack of thugs dragged a couple out of their home and tied them to a tree, where they had a grandstand view of their house being torched. I got many of these incidents on tape, and four or five times that month my pictures led TV3 News.

In broadcasting, TV3 was the new kid on the block – New Zealand's first private channel. It was only two or three years old and just beginning to make an impact on the way television handled news. Television New Zealand, meanwhile, still did things much the way it always had: the TV One newsroom heard about an event and sent a camera crew, which usually arrived after it was all over. If the channel got anything good it screened it, but more often the footage – firemen packing up, or hosing smoking ruins; cars being towed away while people swept up the glass; battered fences showing the height of floodwaters – was too boring to bother with.

Now suddenly, night after night, TV3 News was carrying sequences of fire-fighters frantically deploying as flames rose two storeys high; flashing red lights flickering over twisted car wrecks, with fire and ambulance crews working to rescue people still trapped inside; floodwaters raging through main streets and pouring through houses. The bar had been raised: there was now a big gap between the dramatic footage on TV3 and the boring stuff on TV One.

Television New Zealand had to respond and it did. I got a call one day from its Christchurch bureau chief, offering me a job doing what I was doing for TV3. When I declined, they hired someone to compete with me. TV3 realised it couldn't afford to be upstaged, so it offered to formalise the arrangement between us. Suddenly I was a news cameraman, with a salary and a brand new broadcast camera. I had, to all intents and purposes, pioneered a new industry within the New Zealand media. I had written the job description, and then been offered the job.

Soon afterward, Mark Jennings shifted to TV3's Auckland news bureau. He found there was no one there chasing emergencies the way I did, so he phoned and suggested I shift too. I decided to go. I had been flatting with another TV3 cameraman called Joe Morgan. Joe had accompanied me on a few of my emergency gigs and had caught the bug. He stepped in and took over my job, while I made the move north. Within days I had found a flat in the Auckland suburb of Ellerslie and installed my radio gear. I was looking forward to chasing even bigger and better emergencies. I didn't have to wait long.

Auckland, being 1000 kilometres north of Christchurch, has more tropical and therefore unstable weather. And as well as frequent floods, it offered plenty of fires, murders and car accidents. One of the most bizarre incidents occurred when a man protesting against the use of 1080 poison to kill possums hijacked a helicopter at gunpoint near Mount Tongariro in the central North Island. Eventually the hijacker made the pilot drop him high on the mountain. A major manhunt ensued, involving the police's Armed Offenders Squad. When I rushed to the scene, I realised a brother of the man was waiting anxiously at the bottom of the mountain.

House fires are the stock-in-trade of television news cameramen. Blazes like this fill news slots on 'slow' nights.

It was clear the two men had made contact. If I could interview the fugitive it would be a huge story.

After a few days the brother was given a cellphone, and he climbed up the mountain and gave it to his brother. I hatched a plan: I would go up the mountain in a helicopter, secure the fugitive's guns, do an exclusive interview with him, and then fly him out to the police station. I decided I had to tell Mark Jennings of the plan as it was a bit over the top. Although Jennings was then TV3's New Zealand bureau chief, he was still a reporter at heart and quickly decided to join me. I remember saying to him on the way up in the chopper, 'We are either going to end up dead or get one hell of a story.' We got one hell of a story. The man handed over his guns without any problem. He then sat down and gave a long and sometimes emotional interview where he was often in tears. His love of the land and the wildlife had driven him over the edge.

The police, of course, were not amused, and other media slammed our actions, as was their wont when we got a story they didn't have. The fugitive got two years in jail for hijacking.

Mount Ruapehu Erupts – Central Plateau, New Zealand

In September 1995 Mount Ruapehu, the highest peak in New Zealand's North Island, spectacularly erupted for the first time in 50 years. I evaded the police cordon at night to climb the mountain and take the photos that launched my international career filming extreme natural phenomena.

In 1995 Mother Nature stepped up to send my career in an unexpected direction. With almost no warning, one of the two dormant volcanoes in the middle of the Central Plateau in New Zealand's North Island erupted. Ruapehu is a beautiful mountain, a 2797-metre composite cone, snowcapped year-round, with three ski fields operating on its slopes in the winter. It is part of an ancient volcanic field that includes nearby Lake Taupo, the drowned caldera of a volcano which erupted catastrophically around 185 AD, causing meteorological disturbances recorded as far afield as China and Rome.

While Ruapehu had erupted numerous times in recorded history, it had been fairly quiet during the twentieth century. Its last burst of significant activity had been on Christmas Eve 1953, when a rise in the crater lake had broken the crater wall and sent a lahar down the Whangaehu River, destroying a rail bridge just before the arrival of a train packed with people on their way home or on holiday for Christmas. The train had plunged into the river with the loss of 151 lives. The Tangiwai rail disaster, as it became known, was a reminder that volcanoes didn't need to toss around great plumes of lava to be deadly.

Since then, most of the activity in the field had centred around Mount Ngauruhoe, a peak regarded by most people as a volcano in its own right but actually just the youngest vent of nearby Mount Tongariro. With 45 eruptions in the twentieth century, Ngauruhoe qualifies as one of the world's most active volcanoes. Its last significant outburst had been in 1977. Now it was Ruapehu's turn. On September 18, 1995, almost 50 years to the day since the mountain's last performance, a sharp increase in seismic activity was detected. Then towards the end of the day a large plume of vapour emerged, as magma making its way to the surface encountered the crater lake. Civil Defence issued a level 4 alert, warning that a significant eruption could be imminent. Sure enough, the vapour cloud began to feature ash.

As soon as I heard what was happening, I jumped in my vehicle. It is around four-and-a-half-hours' drive from Auckland to Ruapehu, and I reached the mountain at three in the morning. On the way, I was exposed to one of the main dangers of attending disasters: death or injury through motor accident. In your eagerness to get to a site, it's easy not to pay the road the requisite amount of attention, or to get heavy on the accelerator. I was also battling a totally irrational anxiety that the eruption would fizzle out before I could shoot it. I had to keep reminding myself that I wasn't going to get anything in the way of footage if I didn't actually get there.

MOUNT RUAPEHU	
Location	New Zealand
Type	stratovolcano
Height	2797 metres / 9177 feet
Last eruption	1997

The police had imposed an exclusion zone around the mountain to protect the public. This kind of measure always amuses me; in the West, we're so used to the state telling us what is and isn't safe we've forgotten how to exercise our own judgement. We even need the police to tell us that an erupting volcano is dangerous! In the developing world, the most you'll get is a hand-written cardboard sign beside the road telling you a volcano is erupting up ahead. Most people then obey their instincts and stay away.

In theory, the access roads to the mountain were closed, but if there had been manned roadblocks in place during the day there was no one to be seen in the wee hours of the morning. I drove quietly along with my headlights off, past Chateau Tongariro, the grand hotel at Whakapapa Village on the lower slopes of the mountain, and up the road to the deserted ski field. I parked in the top car park and got ready for the climb. Although I was on the slopes of an erupting volcano, the greatest danger was not hot rocks and boiling lava but cold. A winter ascent of Ruapehu is never something to be undertaken lightly, especially in the dark, so I had all the necessary gear: headlamp, boots, thermal clothing, wet-weather gear, gloves, balaclava, ice axe and crampons. I was carrying sufficient food and water to sustain me on the climb and if, for any reason, I got caught up there. On top of this I had my camera gear – video camera, stills camera, spare tapes, spare battery, spare film, two lenses and a few filters. It was a fair load.

I set off just after 3.30 a.m. to make the five-hour climb to the summit. The first stages were frustrating, tramping beneath chairlifts which, had they been operating, could have whisked me to within 100 metres of my destination. The consolation was that I was almost certainly the only person on the mountain – certainly the only one heading for the summit. It was hot work, trudging up through the snow. Whakapapa ski field had recently experienced a few lean seasons, but this year everything was shaping up beautifully: a nice base and a couple of fresh dumps had set down thick powder snow. Skiing conditions were perfect, apart from the small matter of the volcano erupting. But the snow you want to ski on is not necessarily the snow you want to tramp in. Too soft for crampons, it required an energy-sapping plod. A thick layer of ash lay on top; the only whiteness was in the holes where my feet sank.

At about 8.30 I reached the Whakapapa glacier just below the summit ridge. The mountain was putting on a beautiful show. The eruption cloud, a huge, boiling mass of steam gilded by the morning light and intermittently muddied at its foot by great jets of black ash, soared from the crater into a flawlessly blue sky. There was an acrid, sulphurous smell – the unmistakable odour of geothermal activity. Every now and then a volcanic tremor would make the ground quiver. I could hardly keep the grin off my face. Each ash burst was accompanied by a deep, rolling concussion, which I felt in the soles of my feet and the pit of my stomach.

Ruapehu's crater lake before the 1995 eruption. The lake held up to 10 million cubic metres of acid water. Within days this had been entirely ejected, and lahars poured down the mountain.

I climbed the summit ridge to the Dome Shelter, an emergency retreat for skiers and climbers caught out by the weather. It was in full view of the site of the activity, and right in the impact zone of the volcanic debris. I had to begin to exercise judgement. As well as ash and steam, the volcano was ejecting rocks hundreds of metres into the air. To be hit by one would spell almost certain death. Fortunately, because there was snow on the ground, it was easy to make out the zone in which the rocks were falling. It was like looking at a probability chart, where the density of white pock-marks in the blackened snow represented the likelihood of your dying.

I approached as closely as I dared, unpacked my camera and began filming. As the green 'Record' light winked in the viewfinder, thoughts of Katia and Maurice Krafft, the French volcanologists whose footage of erupting volcanoes I had marvelled over as a schoolchild, came into my mind. I had dreamed of doing what they had done. Now I was.

Ruapehu at dawn as the mountain blows its top. The plume of water and ash soaring hundreds of metres into the sky dramatically dwarfs the ski-lift in the foreground.

Every time there was a larger-than-usual explosion or a plume of ash, I sneaked a quick look to make sure nothing was falling in my area. An occasional rock hit and hissed on the snow at a distance, but nothing seemed seriously threatening. The Kraffts had died filming a volcano. I had no intention of doing the same.

After a while I decided to do a small stunt. I set my video camera, still rolling and framing the eruption, on top of my pack on the ground, plugged in the external microphone, walked around and crouched down in the lens' field of view. 'I'm about as close as it's possible to get to the erupting volcano,' I told the camera, 'and even here, I'm not safe.' The sequence was probably only 50 seconds long. I did it as a personal memento, rather than with a view to submitting it as news footage. Most people like to get themselves in their holiday snaps, and this was just a glorified version. But in terms of my subsequent career, the segment would prove historic.

By ten in the morning, although I still had tape to burn it was time to get back down the mountain. TV3's deadline was looming, and if I didn't get the tape to the nearest big town, Taupo, in time to transmit to Auckland the whole exercise would have been futile. The trip down was uneventful. I arrived back at the truck and after doing a bit more filming I hit the road, getting to Taupo with hardly a minute to spare. As the footage fed through the fibre-optic cable, I sat chafing, then sped back to the mountain. Today, when I have equipment which enables me to broadcast live from events as they happen, it seems crazy that I had to abandon the action for hours in order to file the footage.

By nightfall, when I found somewhere to stay, fatigue really set in, but I felt good. I'd filmed my first erupting volcano and I was confident that for drama and immediacy my footage would beat anyone else's.

My confidence was not misplaced. What I wasn't prepared for, though, was my ugly mug filling the screen on the six o'clock news. It turned out my little piece to camera had galvanised the TV3 editorial crew. In retrospect, this made perfect sense. Shots of anything, even natural disasters, no matter how spectacular or pretty, are static and dead without a human touch. Certainly such material has its place, generally in natural history documentaries where you can afford to be – indeed *want* to be – dispassionate. But news is different. It is all about drama. You can have all the chaos and mayhem you like, but human beings are what make footage dramatic.

This is an unwritten rule of photojournalism. Whether it's still photographs or moving pictures, having a human being in the frame gives the subject scale, and a set of emotional reference points for the viewer. The person – the witness to the event – becomes the story. Putting myself in the frame, with Ruapehu going full-noise in the background, and each boom and burst of ash being accompanied by my reflexive grimaces and nervous glances, brought the whole thing alive. Without intending to, without thinking about it at all, I had crossed the great divide in traditional television news gathering, the time-honoured distinction between cameraman and reporter. It was a recipe for pure television magic. The footage sold around the world.

There were some interesting sequels. Given that there was an exclusion zone around the whole mountain, the sight of me within a couple of hundred metres of the eruption caused consternation among the police and Department of Conservation officials. The real flak, though, came later, when a goon in sandals and walk-shorts climbed the mountain to take pictures and had to be pulled off by helicopter, risking not only his own life but that of the rescue crew. The authorities decided this was my fault. I was, they said, a cowboy.

In fact, though, I had not taken unnecessary risks. I had no desire then, nor have I formed one since, to be flattened by a lava bomb. Of course no one really knows what's going to happen when a volcano is erupting. But who knows what's going to happen at the next bend of the highway? Just because there are speeding idiots on the roads doesn't mean you lock up your car in the garage and give up driving.

Ruapehu kept erupting on and off or another couple of months. Then, the following June, just as everyone was preparing to put it behind them and have a decent ski season, up it went again. I decided to repeat my close-encounter broadcast. While I was kitting up in the car park, a white Land Cruiser with 'POLICE' in blue lettering rolled up. The officer inside wound down his window and looked me up and down. He clearly knew who I was: even if he didn't know my face, he could see the climbing and camera gear I was busy hanging on my person and cramming in my pack, and there weren't too many others in the same business.

'I don't even want to know what you're doing,' he said, and drove off.

This man would be responsible for perhaps the most interesting sequel of all. A couple of years later, the British television production company Granada – best-known for a little soap opera named *Coronation Street* but also one of the biggest makers and suppliers of natural history documentaries to networks such as National Geographic and Discovery Channel – was planning a series of documentaries on volcanic eruptions. A crew arrived in New Zealand looking for quality footage of Ruapehu's recent outburst. Somehow they found their way to the local police station and got talking to the officer in the Land Cruiser, who had been responsible for public safety during the eruption. He told them they'd go a long way before they'd find better footage than the stuff some lunatic from TV3 had shot from the crater rim.

So, thanks to the man whose job it was to stop people like me getting too close to the eruption, I found myself in negotiations with Granada about the footage I had shot too close to the eruption. They were pretty keen to have the film. I casually asked how much they were offering.

'Twenty dollars,' they said.

'Twenty dollars,' I echoed, without enthusiasm.

'US,' they added.

'What – for the whole thing, five minutes, per minute…?'

'No, no, per second,' they replied. 'Twenty US per second.'

I did a quick mental calculation. Fifteen minutes is 900 seconds. Nine hundred seconds at 20 bucks a second was, by my reckoning, 18 grand. US.

'How much?'

It dawned on me right then that if I could get footage like the Ruapehu sequence and sell it for this kind of money, I could finance another, even more adventurous trip. And if I got decent footage on *that* trip, I could sell it to finance the next one. And so on.

And it just kept getting better. Once they'd seen the footage, Granada, too, was mesmerised. They wanted to talk to me about being in their series.

Satellite photograph of Ruapehu's ash plume, June 18, 1996.
LANDCARE RESEARCH NEW ZEALAND

Volcanoes

Vulcanism – Messages from the Underworld

Volcanic eruptions, like earthquakes, tsunamis and mountain ranges themselves, are a by-product of the secret inner life of the planet, better known as plate tectonics. According to this theory the Earth's outermost layer, is divided into a number of plates which are adrift on the next layer, the molten mantle. The plates press and rub up against one another. The plates' boundaries may be *convergent*, with two plates effectively colliding. At this juncture, one plate can be forced underneath the other (subduction) or the two can rear up together along the seam (construction or up-lift). Or the plates may be *divergent*, or drawing apart. In both cases, the plate boundaries are subject to unspeakable stresses and strains. The rock of which the plates are composed tends to fracture in these zones, and one of the possible consequences of fractures is vulcanism – volcanoes.

Beneath the crust, at greater or lesser depths depending on the pressures which are locally present, reservoirs of molten rock – called magma – are formed. Magma is fluid and lighter than solid rock, and can therefore be moved about under pressure. Where a weak point in the crust occurs, magma is forced to the surface by superheated gases, and a volcano is the result. The point where the magma emerges is called a vent.

The type of eruption that occurs depends on the speed at which the magma is moving when it emerges. Very fast-moving magmatic material

Mount Etna's Voragine crater erupting on the night of June 26, 1998. Part of Etna's central crater, Voragine formed in 1945, and had grown progressively larger with strong eruptions from the 1960s onwards, reaching a diameter of 350 metres.

is blasted into the air and may disintegrate into larger and smaller particles, generally called tephra. Small, dust-like volcanic material is referred to as ash, and this may settle some distance from the site of its emergence. Larger, shingly material – cinders – usually rains down locally. Larger bits again, emerging still more slowly, range in size from small fist-sized 'bombs' to 'blocks' of rock as big as buses. A mixture of heavy and light material may drift down the mountainside in a *nuée ardente* (literally a 'glowing cloud'), or a quantity of semi-molten rocks may tumble down in a pyroclastic flow. The slowest-moving material of all may form rivers of molten rock – lava – on the surface.

TYPES OF VOLCANOES

Volcanoes can take a number of different forms, generally depending on what type of eruptions generated them. They are divided into four broad types:

CINDER CONES As the name suggests, these volcanoes form around a vent from which a plume of cinders has been ejected. Over time, the successive falls of ash and cinders build up in a conical heap around the vent. The result is (typically) a low, generally symmetrical mountain with a crater at the summit – a classic volcano.

COMPOSITE VOLCANOES, OR STRATOVOLCANOES When, as commonly happens at convergent plate boundaries, a range of activity occurs at a volcanic site over time, with the speed of the pyroclastic flows varying at each event, a composite volcano is formed. Layers of lava flows, cinder- and ashfalls, studded with blocks and bombs, build up, resulting in a large, more or less regular mountain with a central crater containing one or many more than one vents. These vents connect to a conduit, which leads to the magma reservoir. When volcanic activity eventually slows and ceases, the molten material in the conduit solidifies. This tends to be much harder than the stuff that the cone is made of, and over thousands of years following the cessation of activity the cone itself will weather and erode, leaving the 'plug' exposed.

SHIELD VOLCANOES Where magma slops sluggishly to the surface, and oozes from a fissure or vent, it progressively spreads over the immediate area, forming a low, flat mountain, usually called a 'shield' volcano owing to its resemblance to that piece of armoury. Sometimes the mass associated with a shield volcano may spread over a huge area, forming a great, flat plateau. This type of volcano is most common at divergent plate boundaries, most of which are under the oceans. Some of the better-known shield volcanoes are in Hawai'i, and while they are relatively small as they meet the eye, the vast proportion of their bulk is beneath the ocean.

LAVA DOMES Commonly when a volcanic vent is almost wholly obstructed, material may be extruded through whatever opening is available, forming a dome of molten material crusted over with solidified rock. As the dome increases in size the solid material may break or crumble off. Lava domes are often trying to tell us something, and what they're trying to tell us is 'Run': they can be an early warning of a catastrophic eruption, because they're a pretty good indication that there is a mass of unrelieved pressure beneath.

World's Deadliest Volcanic Eruptions

VOLCANO	YEAR	MAJOR CAUSE OF DEATH	FATALITIES
Tambora, Indonesia	1815	starvation	92,000
Krakatau, Indonesia	1883	tsunami	36,417
Pelée, Martinique	1902	ash flows	29,025
Ruiz, Colombia	1985	mud flows	25,000
Unzen, Japan	1792	volcano collapse, tsunami	14,300
Laki, Iceland	1783	starvation	9350
Kelut, Indonesia	1919	mud flows	5110
Galunggung, Indonesia	1882	mud flows	4011
Vesuvius, Italy	1631	mud flows, lava flows	3500
Vesuvius, Italy	79 AD	ash flows and falls	3360
Papandayan, Indonesia	1772	ash flows	2957
Lamington, Papua New Guinea	1951	ash flows	2942
El Chichón, Mexico	1982	ash flows	2000
Soufrière, St Vincent	1902	ash flows	1680
Oshima, Japan	1741	tsunami	1475
Asama, Japan	1783	ash flows, mud flows	1377
Taal, Philippines	1911	ash flows	1335
Mayon, Philippines	1814	mud flows	1200
Agung, Indonesia	1963	1963	1184
Cotopaxi, Ecuador	1877	1877	1000
Pinatubo, Philippines	1991	1991	800
Komagatake, Japan	1640	1640	700
Ruiz, Colombia	1845	1845	700
Hibok-Hibok, Philippines	1951	1951	500

Mount Etna, photographed from the summit of its south-east crater, Voragine, during an eruption on June 26, 1998. The city of Catania, threatened for centuries by Etna's activity, can be seen in the distance

The Greatest Show on Earth – Mount Etna, Italy & Mount Stromboli, Italy

One of the accidental but happy consequences of having done a piece to camera during the Ruapehu eruption was that the sequence served as a superior kind of business card. The footage went around the world, and my face got most of the airtime, which was not a bad way to draw attention to myself and my work. One of those who came knocking was Pioneer Productions, which had an idea for a series of documentaries called *Volcano Detectives*. Discovery Channel would screen the series, and they flew me to London to talk about it.

The series, which we filmed in 1998, was an opportunity to travel to some of the volcanoes I'd only read about, or seen on television. Via the internet I had met John Seach, an Australian food technologist who moonlighted as a volcanologist and tour guide to Earth's hot spots, and I suggested to Pioneer that he help on the show. We sat down and drew up a wish list. As it grew to include names such as Krakatau, Etna and Yasur, my mouth started watering.

MOUNT ETNA	
Location	Italy
Type	stratovolcano
Height	3323 metres / 10,902 feet
Last eruption	2005

Our first port of call, as it would be for anyone in search of the greatest show on Earth, was Mount Etna in Sicily. We flew into Palermo in the middle of summer. If we were looking for danger, we found it right away: Italian driving. Drivers seemed to start, stop and swerve with no signal or warning, and the whole mess of cars, vans and trucks was threaded by lunatics on Vespas and pedestrians with a death wish. We only just survived the taxi ride from the airport and the 230-kilometre drive to the town of Taormina, at the foot of Etna, on Sicily's east coast.

In Taormina it's hard to ignore Etna. The volcano's smoke plume frequently blots out the sun, and there's a fine coating of ash on everything most of the time. With the larger tremors, the ground shakes and glasses tinkle in their cupboards. The view at night, with the glow from the peak and even the odd skyrocket of magma – has to be seen to be believed. Our brief was to illustrate the majesty and destructive power of volcanoes, covering both historical and contemporary eruptions. I spent the next day filming around Etna's feet, and there was plenty to shoot, including villas and farm buildings which had been abandoned, destroyed or half-buried during Etna's larger tantrums.

New Zealanders are reasonably relaxed about volcanoes, even though we live smack bang on the rim of fire, the band of volcanic activity which girdles the globe from Hawai'i through Southeast Asia and down to Antarctica. Auckland, our largest city, is built on a volcanic field featuring over 100 cones. But we have nothing on the Sicilians, who have built right up to the feet of Europe's largest volcano, which also happens to be one of the world's most active. There are four major towns within 50 kilometres of the caldera, and any number of villages, farms, vineyards and houses, some built in defiance of laws designed to protect people from the volcano. In fact, Sicilians have even built a ski field and a gondola on the cone itself, although every time activity increases they get wiped out.

Most of the world's volcanoes occur in zones of tectonic subduction, where one of two convergent plates is forced beneath the other. Etna is unusual in that it lies on the constructive boundary – that is, where the plates crumple upward together – between the African and European continental plates. This geological uplift will one day obliterate the Mediterranean Sea. The pressure creates faults, through which magma forces its way. Etna was formed 500,000 years ago on one of these faults and in its early years was the site of mammoth, fissure-type eruptions. These gradually centralised and a shield-type cone developed. As long as people have lived in Sicily, Etna has been there, grumbling away and occasionally spitting the dummy – as in the great eruptions of 475 BC and 396 BC.

A large lava fountain shoots from Mount Etna's south-east crater. Photographed from the town of Nicolosi on June 12, 2000.

One of the most ferocious eruptions of more recent times took place in 1669. Great lava flows began heading for the towns of Nicosia and Catania. To save Catania, ditches and earth walls were thrown up. However the effect was to increase the threat to Nicosia. The works were reversed, Catania was destroyed, so too was Nicosia, and the lava flows reached the sea. In the wake of this disaster, legislation was passed prohibiting attempts to divert lava flows. This stood until 1983, when it was repealed to allow work to be done to divert a flow which was threatening Nicolosi and Belpasso. Again, channels and earth dams were constructed, but these were only partially successful.

Since that time, Etna has had thousands more eruptions, including four major ones in the nineteenth century and eight in the twentieth. In 1992 American helicopters were roped in to drop huge concrete blocks into a vent which was the source of a flow threatening Zafferana. This too, while enterprising, was only a qualified success, and you wouldn't recommend it in Etna's really bad moods. Some eruptions have been long. Two have consisted of a year or more of continuous activity. But Catania and Nicosia are still there, rebuilt and inhabited by cheery Sicilians as though nothing has ever happened and nothing ever will. In fact Etna is one of Sicily's greatest assets, a huge tourist drawcard.

In recorded history only 73 people have died as the result of Etna's activity. John and I kept reminding ourselves of this as we shouldered our packs and began the long, weary walk to the summit. Climbing any volcano is tedious: the cones are deep in ash and cinders, and each step forward results in a slide which seems to take you two steps backwards. It's like walking on ball-bearings. And it was hot. The volcanic material was absorbing and reflecting the Mediterranean sun like crazy.

Our destination was a ridge above the Bocca Nuova crater, which was the scene of some spectacular Strombolian-type activity, with pyroclastic spurts regularly throwing material 300 metres into the air. We arrived before sunset and set up our gear to film the fireworks. The show was spectacular enough in daylight, but even more incredible at night. What we were after was a shot to illustrate the threat Etna posed to the surrounding inhabited areas, and we got it: lava being tossed into the air, with the lights of Taormina shimmering in the heat-miraged background.

The crater was a humbling place to be. Each incandescent ejection was accompanied by a bone-shaking report, like an artillery round, and underlying it was the roar of superheated gases escaping from vents and fissures around the caldera. Every now and then we'd feel a strong tremor, and wonder whether that meant an escalation of activity. Etna's caldera is a large, unstable area, with new vents liable to be formed any time, anywhere. We were on what seemed to be safe ground, but we knew there were no guarantees. Most of the 73 deaths had been the result of people approaching the active area too closely. However I had a job to do, and the shots I was getting – both with my digital video camera and stills camera – were enough to keep my mind off the possibility that the next active vent on Etna could open up where we were standing. We kept rolling for a little under three hours. Then we made our way, under headlamps, back down the mountain.

The island of Stromboli as we approach it by hydrofoil, June 28, 1998. The island is simply the summit of a volcano, most of which lies beneath the sea. It is one of the most active volcanoes on Earth.

Strombolian activity is named for the volcanic island of Stromboli, which lies around 75 kilometres north of Sicily and is part of the same volcanic fault as Etna. This was the next stop on my itinerary. I travelled by launch – or, more accurately, hydrofoil – from Tropea on the Italian mainland to the little settlement of San Vincenzo, another town built in the shadow of volcanic destruction. There's not much to Stromboli other than the volcano, which has been in almost continuous eruption for the last 2000 or 5000 years, depending whom you talk to. Everyone who goes to Stromboli goes to look at the volcano, and every one of the 450 or so people who live there is involved in the volcanic tourist trade, one way or another. Soon after I arrived, I met up with a party of tourists: two British women, an Australian, a Frenchman and a fellow New Zealander. They were impressed with my Etna stories – a couple had seen the volcano themselves, but not as close-up and personal as I had – and asked whether they could tag along on my climb to Stromboli's crater. Well, why not?

MOUNT STROMBOLI	
Location	Italy
Type	stratovolcano
Height	924 metres / 3031 feet
Last eruption	2006

34

At only 900 metres, Stromboli is nowhere near the climb Etna is, but it was another glorious Mediterranean summer's day and ferociously hot. Even so, we did it at a cruise. My plan, as with Etna, was to film the lava plumes at night. We found a likely-looking vantage point, laid out our sleeping bags and set up our cameras. I filmed the sun sinking over the Mediterranean, with a cargo ship making its way along the horizon. It was unbelievably beautiful. That kind of postcard shot, though, had nothing on what I got from Stromboli. Over the few hours I filmed, the level of activity gradually increased. At first there were some loud bangs, and glowing material was hurled from the vent. Once the sun had set, though, the volume of both the noise and ejected material grew. There was a pattern to it: two or three moderate bangs would be followed by a pause, then a much bigger explosion. At one point, as a pause lengthened, I observed that it seemed to be building up to the big...

'Bang!'

'That,' I said to the British women, 'is better than sex, isn't it?'

By now some of the lava bombs were falling just short of our position. I wasn't seriously concerned, but I have to admit I didn't get much sleep, what with the hard ground, the enormous explosions and the earth moving beneath us. The French tourist had brought along a kind of metal screen, such as you would see used on a barbecue. He put this over his head before he went to sleep. What difference it would have made if he had been hit by a car-sized rock I don't know.

Stromboli has been in continuous eruption since Roman times and shows no sign of stopping. I took this photo from the summit area looking down on one of the active craters during my 1998 visit. In December 2002, for the first time in 17 years, lava flows reached the sea, causing panic among the island's normally stoic inhabitants, and the summit was placed off limits.
In 2006, as this book was in production, the volcano was again highly active, sending 4000-metre lava plumes into the air.

What better way, I decided, to round off my Italian jaunt than by visiting the world's most dangerous volcano, Vesuvius, and the remains of the two victims of its 79 AD hissy fit, the Roman towns of Herculaneum and Pompeii. Vesuvius looms over the Bay of Naples and is well over 1000 metres high, but apart from an occasional wisp of steam from the summit, the monster was quiet for our visit. I felt strangely moved, walking about the ruins with the dark, cocked silhouette of the mountain brooding in the background. We climbed to the crater, but apart from some fumarole activity there wasn't much to see.

RIGHT
The crater of Vesuvius looks calm these days but it is famous for a massive eruption in 79 AD which buried the towns of Pompeii and Herculaneum under metres of ash and volcanic debris. The mountain has erupted over 30 times since, most recently from 1913 to 1944.

OPPOSITE
Daybreak at the summit area of Mount Stromboli. Nearly 500 people live in the shadow of the highly active volcano.

John and I next visited Etna in the winter of 1999. By then the mountain had upped the ante, producing further intense Strombolian activity at the Bocca Nuova crater, including from a vent close to the spot where we had filmed the year before. In November we ascended to the summit past signs which read, in Italian and English: 'It is dangerous to go beyond this limit due to volcanic activity.' But the greatest danger was the cold. For much of the time, the temperature was minus 15°C (5°F), and occasionally even colder.

We camped overnight about a kilometre from the crater, but with the cold and the ear-shattering explosions neither of us got a hell of a lot of sleep. The reward, though, was some memorable footage. In one of my particular favourites, a relaxed-looking John Seach is half-sitting, half-reclining on a low mound of slag while lumps of glowing material leap up from the crater rim behind him. He's a picture of nonchalance. Suddenly there's a huge bang, and a great bloom of glowing orange rocks appears behind him. His head snaps around and he falls off his perch. It began to look as though Etna had lit up another crater right behind our campsite. Lava bombs half a metre in diameter and easily big enough to kill us – fireman's helmet or no fireman's helmet – began landing nearby. We looked around for a position free of impact craters: having come all the way up and spent a freezing night, it seemed a waste to have to head down. But as material kept raining down all around us and the mountain issued ever more explosive reports, we agreed it was time to go.

This episode marked the beginning of a sustained series of vigorous eruptions – over 60 in the next four and a half months. We were there in June 2000 for the climax. We had already had an exciting time chasing volcanoes that year, but nothing we'd seen in Vanuatu, Indonesia or even Japan had prepared us for the display we were about to witness. June 8 found us camping in a stone bunker,

TOP
Strombolian activity at Mount Etna's Bocca Nuova crater, November 14, 1999, with volcanologist John Seach taking a close look. Active volcanoes are constantly changing shape. The Bocca Nuova – 'new mouth' – formed in the western part of the central crater in 1968.

BOTTOM
This violent strombolian explosion at Bocca Nuova sent John Seach (pictured) and I fleeing for our lives as large lava bombs landed within metres. The image was captured by a television camera, retrieved intact after the explosion.

which had formerly housed boiler equipment for the ski field, about a kilometre from the latest crater to explode. The volcano was firing huge plumes of intensely glowing lava 800 metres straight up into the air. Think of a burst water main, scale it up 200 times, and for water substitute red-hot lava, and you have some idea of what it looked like. What it felt like is beyond my powers of description. The noise was like a cross between that of a hurricane and an earthquake, and it never stopped.

This was a better than once-in-a-lifetime photo opportunity. After we set up our gear, John and I took turns photographing each other in our silver heat suits, standing on a ridge while the volcano went full blast in the background. Even in a full heat suit with a helmet and an armoured glass visor, all rated to 1000°C (1832°F), the heat was intense.

During Etna's eruption on June 8, 2000, John Seach watches a 750-metre lava fountain from Etna's south-east crater. It lasted 23 minutes and provided the most spectacular video footage I have ever captured.

Then, without warning, something changed down in the crater and the plume leaned towards us. We were suddenly in real jeopardy. A rain of cinders began falling on us, and an occasional bomb. The chunky bits got more frequent, and more chunky. We rapidly retreated to the comparative safety of the bunker. Rocks hammered on the iron roof, and a fist-sized bomb dropped through a hole and struck John on the upper arm. Lucky for him it was a glancing blow, and not on the head. We had brought enough food and water to last a few days if we got stuck and someone had to come and dig us out. For a while, that looked a strong possibility.

Overnight and during the next day, however, the activity decreased. The lava fountain subsided into a dirty cloud of ash, and finally into a column of boiling white vapour, occasionally stained with ash. We stuck around and kept filming, and weren't we glad we did. All of a sudden there appeared, lazily lifting above the main vapour column, a glowing ring of smoke, dragging a gauzy trail behind it. Neither of us had ever seen anything like it before. We filmed the massive smoke ring, then another, and another, as they rose like jellyfish into the blue sky. Later we learnt that such rings arise from a volcano-induced version of a tornado: a little ball of superheated gas belches from a vent, forming a bubble of low air pressure, which rises, wreathed in visible smoke. Only one vent in Bocca Nuova is known to produce these rings, and then only rarely. The last recorded occurrence had been in 1968, and this was the first time the phenomenon had been caught on camera, let alone film. We felt privileged to have witnessed it – a fitting finale to the greatest show on Earth, executed with distinctly Italian flair.

OPPOSITE
Smoke rings coming from Mount Etna's Bocca Nuova crater. These events are extremely rare. Before we saw this in June 2000, one had not been recorded for 30 years.

In April 2000, volcanologist John Seach and I evaded cordons around the town of Toyako Onsen on the northern Japanese island of Hokkaido and camped in the town for three days, until driven out by lava flows. Here the action begins.

Danger on Main Street – Mount Yasur, Vanuatu & Mount Usu, Japan

In 2000 John Seach and I teamed up again for another round of filming for **Volcano Detectives**. *In February we flew to Vanuatu, and then on to Tanna in the southern part of the island group. Tanna hasn't much going for it in the way of tourist assets, but it does have Yasur volcano, one of the world's most active.* The first Westerner reported to have seen Yasur was Captain James Cook, who in 1774 spotted its glow from afar. He asked the local inhabitants for permission to climb the sacred mountain but was refused.

A 361-metre stratovolcano which is easily reached and puts on a pretty show, Yasur every year draws many volcanologists. We had an adventure in mind: we were going to abseil into the crater and film the action at close range.

You can approach Yasur a number of ways, including by helicopter. We chose to cross the wide, blasted cinder plain surrounding it by mountain bike but quietly, off-camera, our gear was advanced to the foot of the cone by car and dumped in the car park. Of course, there was no alternative to climbing the cone itself on foot, lugging our gear on our backs, so that bit at least was as arduous as it looked on the show.

We camped on the summit for three days. The plan was to drop into the caldera on its south-east edge. There are several active craters within the caldera itself: while we were there Crater C was most active, ejecting magmatic material and loudly degassing. It's actually not a big deal abseiling into volcanic craters. When you mention it to people they get a picture of huge heaving cauldrons and enormous fountains of lava. A bit of that goes on, but with luck it's at a safe distance. While we were on the summit ridge we were lucky enough to witness and film two lava plumes being tossed into the air – which added to the drama of what we were about to attempt, as witnessed by someone on their living-room couch. We noted that most of the material was going straight into the air or away from our angle of approach.

MOUNT YASUR	
Location	Tanna, Vanuatu
Type	stratovolcano
Height	361 metres / 1184 feet
Last eruption	2006 (ongoing)

The greatest danger with a stunt like this is not, curiously, incineration, or even being hit by a flying rock, although both threats exist. There's a slightly higher danger of being asphyxiated by toxic gas, although this is easily countered by wearing gas masks. The real danger is the workaday one you face climbing on any unstable face: rockfall.

The descent into the crater on February 18 was routine. On touching down on the level ledge adjacent to the active vent, wearing his silver fire-fighter's suit and helmet, John became the first person to have abseiled into Yasur. This was the real point of the exercise. Although John carried a trenching shovel and dug desultorily in the ash, there was no real scientific value. It was interesting to note, however, that the ground temperature where he was standing, about 15 metres from the edge of Crater C, was 95°C (203°F). With a telephoto lens, the distance between John and the occasional jet of lava was foreshortened. All the same, when you watch the film and see him flinch, it's authentic.

This photo of Yasur volcano, which I took from the air in December 2003, shows its huge, 400-metre-wide crater, which forms a nearly perfect circle. Yasur may have been erupting continuously for as long as 800 years.

There were some seriously loud explosions and a lot of glowing material hurtling around. Shortly after he had beat a retreat, there was a louder than usual bang, and bombs descended close to where he had been standing.

From Tanna we flew to back to Port Vila, and then hopped on a chartered single-engine Cessna to the island of Epi. From here we took a boat to another island volcano. Lopevi is a dramatic mountain, rising straight out of the sea to a height of nearly 1500 metres. It has erupted over 20 times since 1863 and its inhabitants have long since moved to nearby islands to live.

At Epi, we had learned that Lopevi had erupted while we were on Yasur. Two loud explosions had occurred on the night of February 19, and a lava flow had emerged from a vent on the north-west flank. When we reached the island, we had no difficulty finding evidence of this activity. The pyroclastic flow, 300 metres wide, had reached the sea 2.5 kilometres from the vent where it had issued, cutting a swathe through the rain forest on the island's lower slopes. Some of the trees were still burning, and the air was full of the stench of burned vegetation and sulphur. Every now and then a superheated coconut would explode, adding another hazard to our list of things to watch for when volcano-chasing in the tropics. We poked about Lopevi for a couple of days – climbing to view the site of the eruption and walking about on the lava flow – until our boots melted and we were forced to retreat. It was all good material for the show.

From Vanuatu the plan was to visit two Philippines' hot spots: Pinatubo, which famously erupted in 1991, the second largest eruption of the twentieth century, killing 740 people and disrupting the lives of two million more; and the beautiful, classically shaped Mayon. I had a better idea though: monitoring the internet I'd noticed that an alert had been issued in the vicinity of Usu, a volcano on south-west Hokkaido, Japan's second-largest island. Swarms of earthquakes had been hitting the area – some tremors which could be felt, and other low-frequency shakes which could not but which pointed to big subterranean movement. Something was going to happen.

Even on the map, Usu is a good spot for volcanology. It's situated on the shores of Lake Toya, which is the caldera lake of a sizeable Pleistocene volcano. Both Toya and Usu are a little inland from the evocatively named Volcano Bay, a wide arc which suggests the bay itself is yet another ancient volcanic caldera. Usu, then, is just the latest incarnation of a very old volcano indeed. At the foot of the cone is a small resort town called Toyako Onsen, the *raison d'être* of which is a hot spring, further evidence of geological hot spots not far below the surface.

Usu is a stratovolcano, with two main crater areas and a number of vents. It has erupted, with considerable violence, nine times since 1626: five of these eruptions have caused loss of life through a range of effects – pyroclastic flows, ashfall and lahars. There were three events in the twentieth century: a big one in 1910, another in 1944–5, and a protracted series of eruptions from 1977 to 1982. This time the earthquakes, and the evidence of magma forcing itself up through surrounding rock, closely resembled the action that had taken place before the big 1910 eruption.

On the evening of March 28, Japan's National Coordination Committee of Volcanic Eruption Prediction declared an eruption of Usu to be imminent. Around 1600 people staying at the spa resort at the mountain's foot and another 9000 people from the town were evacuated, and a ten-kilometre exclusion zone imposed. On April 1 a shake of magnitude 4.8 seemed to indicate things were escalating. New vents were appearing, some alarmingly close to the town. They were producing geyser-like eruptions, with jet-black material shooting high into the air. Pieces of pumice found floating in Lake Toya suggested the presence of magma, and regular surveys showed the whole north-western side of the cone was swelling and stretching.

By the time John and I arrived on April 3, things were looking pretty good. That day 22 earthquakes had been felt, on top of 48 the day before. Intensive surveying had shown impressive uplift of the existing crater areas, consistent with the formation of a non-visible lava dome, or cryptodome, and the crater walls were showing signs of fracturing. A couple of significant new faults had been identified. Each time the volcano was overflown, new craters showed up, and their plumes of steam were featuring more and more greyish ash. Small lahars were happening regularly: as the volcanic activity melted snow and ice, the meltwater raced downhill, carrying mud and debris with it.

A media scrum had developed at the roadblock set up by police at the limit of the exclusion zone. Dozens of camera crews and photographers, mostly Japanese, were swarming around, politely observing the authorities' instructions not to go beyond the cordon. We waited until it got dark and then, with all our camera and camping gear stuffed in, strapped on or otherwise attached to our packs, we slipped past the roadblock and set out on foot for the town. It was a ten-kilometre walk on an icy road. In the depths of the Japanese winter, the temperature was minus 10°C (50°F). All the way we could hear the volcano rumbling and hissing, and the occasional flicker from the vents. The smell of sulphur seemed intensified by the cold.

MOUNT USU	
Location	Hokkaido, Japan
Type	stratovolcano
Height	737 metres / 2418 feet
Last eruption	2000

Dawn found us in the town of Toyako Onsen. The asphalt on the roads was cracked and everything was covered in a thick layer of ash. We chose a concrete multi-storey building and made ourselves comfortable. This was where we would sit and watch as the volcano blew its top and took the town to pieces.

At first we felt comparatively safe. We noted that volcanic bombs were falling 200 metres away and, as I explained to camera, 100 metres was our limit. If the mountain started throwing debris closer than that, we were out of there. Of course, you can calculate only so much of the danger a volcano poses. If the eruptive activity were to escalate, there ought to be plenty of warning – increased tremors, explosive shocks and, of course, the pitter-patter of pyroclastic material dropping around us. Exactly what we should do if things quickly escalated was another matter. The two incalculables were lahars from the summit and tsunamis from the lake. The first were likely: the cone was blanketed with snow and ice, which was melting in contact with magmatic material. And there was some precedent in earlier eruptions for the second.

We were almost literally between the devil and the deep blue sea.

Lava surges down the main street of the Japanese resort town of Toyako Onsen as Mount Usu blows its top. All residents had been evacuated.

We filmed the progressively intensifying eruption on the slopes above the town for three days. On day three, it looked as though things were going to get hairy all right. Overnight, a new vent had opened up in the main street a few blocks away, and was ejecting plumes of steam occasionally shaded grey. This was not good. Then, as we were filming, a sudden rush of grey, steaming water slid through the street right in front of us. Enter the lahar. The water, superheated by the volcano, was dragging mud with it and moving at terrifying speed. It wouldn't be possible to outrun it, and if we were caught in it we would be boiled alive.

Understandably, we were anxious to get our gear together and get out in a hurry. But we were also conscious that this was a rare opportunity. Not many people get so close to volcanic activity and live to tell the tale, let alone capture it on film. We were packing up but still filming when the second lahar arrived. For all their power and the violence of their origins, lahars are deadly quiet. The first we knew of it was a quiet hiss of steam behind us. An inch or so of mud and water was flowing across the road. Of course I filmed that, too, but our situation was now precarious. We were in imminent danger of being caught between the two flows. Then we would be at the mountain's mercy.

Grabbing our gear, we started the long hike out. There was an apprehensive prickle between my shoulder-blades as I turned my back on a landscape that was dramatically coming apart. The further away we got, the happier we felt. We were lucky: we had managed to retreat before retreat became impossible.

The Japanese authorities were not happy when they saw us appear from the direction of the town, but what were they going to do about it? The assembled media treated us with a mixture of bewilderment and awe. Every one of them had been just as keen as us to get the best shot possible; it just never occurred to them that you might disobey an official instruction. Although I had no trouble selling my footage to networks worldwide, the Japanese media wouldn't touch it. As far as I know, to this day absolutely none has ever screened on Japanese television: it would send the wrong message to show something that was obtained through disrespect of authority.

After our escapade, we stayed around to film Usu from an observatory situated on a nearby hillside. There's something hypnotic about watching molten rock, especially a slow tumble of the smoking, black variety which breaks open every now and then to reveal its intensely glowing core. At the observatory, I also met volcano-chasers who were familiar to me through the internet, including Isamu Kitafuji, a worker at Tokyo's Narita Airport who chased and filmed volcanoes for a hobby. (In 2002 Isamu would join me in a frustrating jaunt to Mount Karymsky on the Kamchatka Peninsula, the part of mainland Russia which dangles toward Japan. We were fogged in for a week while the volcano erupted unseen.)

When all the excitement of Usu had died down, John and I visited two of the many, many active volcanoes in Indonesia, travelling first to Java's Kawah Ijen, literally the 'crater lake of Kawah' and best known for its sulphur mining. Molten sulphur issues from several vents around the shores of the lake, glowing orange until it cools to its more familiar brilliant yellow. There was a large eruption of pure molten sulphur here in 1952.

Java's Kawah Ijen produces a rich harvest of volcanic sulphur. Here a worker carries a typical load – up to 100 kilograms at a time.

In these quieter times, there's a whole industry based on the extraction of rock sulphur, but it's not especially lucrative – or at least, not for those at the sulphur-face. We filmed workers hacking chunks of the yellow rock, and putting it into baskets dangling from either end of a long pole. One of these loads might comprise a 100-kilogram dead weight. The worker would heft the pole and balance it over his shoulder, then climb out of the crater and down the slopes. To say this is hard work is understatement on a grand scale.

And it's all carried out in what could be called an occupational health and safety inspector's worst nightmare. As with any volcano, the slopes are a mixture of sliding, rolling cinders studded with boulders of razor-sharp scoria. Java gets warm at the best of times, but in the crater area it's the best part of

50°C (122°F). The air contains hydrogen sulphide and sulphur dioxide and monoxide, which form sulphuric acid and sulphonic acid on contact with water. If you get close, your eyes sting, your lungs burn and your skin alternately itches and stings. And if the volcano silently steps up its gaseous emissions, you asphyxiate.

On top of this, there is molten sulphur flying around, and the nearby crater lake, with a pH of 0.7, is one of the most acidic liquids you'll find outside a car battery. As for protective clothing, most of the men we saw were wearing sandals, jeans and sweatshirts, with only a damp rag over their mouth and nose as protection against gas. Danger money? Each trip to the summit and back down with a 100-kilogram load nets the worker US 80 cents. If you're fit and don't suffer any mishaps, you may manage three trips in a day.

I usually feel pretty good about what I do for a living. I came away from Kawah Ijen feeling extremely fortunate as well.

The logical way to round off a trip like this was with a visit to the legendary Anak Krakatau, an 813-metre volcanic island brooding out in the strait between Java and Sumatra. Better known in the West as Krakatoa, Anak Krakatau holds the record for the loudest eruption in recorded history.

While the mountain has been continuously erupting for centuries, the big one came in 1883, when a vent went up catastrophically, totally destroying the cone – only a shattered remnant remains – and generating a 40-metre tsunami which impacted right around the region and the margins of the Pacific and Indian Oceans, as far afield as Africa and South America. While the exact figure will never be known – many tiny isolated communities would have been wiped out, with no one and nothing left to testify to their existence – at least 36,000 people died.

The cone that has been the site of the volcano's activity since that event had been quiet for two days when we visited and climbed it, but Anak Krakatau remains one of the most dangerous volcanoes on Earth to walk on. You didn't have to experience any activity at all to get a thrill from being there.

A large rock ejected from Krakatau's crater. After the renowned 1883 eruption – which generated a tsunami recorded as far away as 7000 kilometres – the mountain was quiet for 44 years. However, since 1927 there have been frequent eruptions.

Misguided Misadventures – Mount Merapi, Indonesia Mount Arenal, Costa Rica & Mount Cotopaxi, Ecuador

In 2001 I went on another volcanic odyssey. This began with a trip to Indonesia to watch Mount Merapi, a 2800-metre stratovolcano in Eastern Java, erupt. Or should I say try to watch it? I arrived on January 21 and survived the traffic in Jakarta and Yogyakarta, and the drive to the volcano observatory at Babadan, only to find the mountain clagged in with low cloud. Bad weather persisted for the better part of three days, allowing only glimpses of what otherwise had all the appearances of a spectacular eruption. But on January 24, with providential good timing – the rest of the international media scrum had packed up and left only an hour before, complaining about the weather – the cloud rolled back to reveal the volcano in full eruption in perfect conditions. At that point I was sharing the observatory with about a hundred people, all of whom were gassing away more voluminously than the mountain; their chatter was recorded for posterity.

MOUNT MERAPI	
Location	Java, Indonesia
Type	stratovolcano
Height	2914 metres / 9560 feet
Last eruption	2006

At about 1 p.m. local time, I was wandering past the seismograph when I happened to notice the stylus scratching big thick lines on the drum. I quickly set up to film, and waited patiently while thick cloud blew through the summit area. I expected a pyroclastic flow to emerge, and sure enough one did – the biggest yet. It headed right for the observatory at motorway speed.

There was pandemonium. Loud hailers blared, people ran to and fro, and soon I was the only person left, still calmly filming. I was relying on the underground bunker beneath the observatory to save me if the flow arrived. This appeared a better bet than trying to outrun the flow, which everyone else seemed to be trying to do, carrying their babies in their arms as they sprinted down the road. Fortunately, a ridge stopped the flow about a kilometre distant.

Lava avalanches pour down erupting Mount Merapi, seen from Babadan Observatory on January 25, 2001. Merapi is one of the most dangerous volcanoes on the Pacific rim of fire. In 2006 it would again hit world headlines when local villagers refused to leave their homes despite the mountain erupting violently. Two men were killed when an emergency shelter failed to protect them from the searing heat.

The good weather lasted into the night, which is much the best time to film pyroclastic flows: the smoke and vapour which normally veil the solid material become transparent and all you see is the incandescent avalanche of rocks tumbling down the slope. By 5.45 in the morning it seemed to be all over. After a few hours cloud again veiled the cone, obscuring the biggest flow of the day, which made it all the way down into the rain forest at the mountain's foot. But the weather was on a downward slide. And Cyclone Terri, brewing in the Coral Sea, was calling.

My big trip of 2001, designed to be a prelude to the tornado-chasing season in North America, was the fulfilment of a lifelong dream to visit the significant volcanoes of Central America. In Costa Rica, my first stop, the country's youngest stratovolcano, 1657-metre Arenal, has been in a state of continuous eruption since 1968, after awaking from a 500-year dormancy. Arenal is a little over 100 kilometres north-west of the capital, San José. I travelled there with Isamu Kitafuji. This was his eleventh visit; he assured me that the spectacular cascade of pyroclastic material we witnessed on the night of our arrival, March 30, was nothing compared to what Arenal could turn on.

MOUNT ARENAL	
Location	Costa Rica
Type	stratovolcano
Height	1657 metres / 5436 feet
Last eruption	2006 (ongoing)

The safest way to get to the top of any dangerous mountain, especially a volcano, is usually to take advice from the local people. In deference to this rule, I hired a guide to take me to the top of Arenal. The experience that followed went to show there was an exception to every rule.

The day started at 6 a.m. in heavy fog and rain. The guide turned up as arranged. He had no food or water, no backpack, no hard hat and was wearing shorts, a T-shirt and bare feet. When I pointed to his feet, he triumphantly produced a pair of gumboots from a string bag. I had 20 kilograms of gear for him to carry. If a shopping bag was his choice to carry it in, well, that was up to him.

Climbing Arenal is prohibited, so my guide had devised a novel way around the official cordon. We plunged into the jungle about five kilometres from the safe, easy point of entry, which the police happened to be watching. We hadn't been going very long when I heard an unsettling noise, half cough, half roar. My guide shrugged. Only a jaguar. *Only a jaguar?* Hadn't I read somewhere that a jaguar would not hesitate to attack anything – human beings included – which crossed its path?

ABOVE
Mount Arenal in Costa Rica was considered dormant until it suddenly and violently erupted in July 1968, burying three villages, killing 87 people and devastating over 200 square kilometres of countryside. Today, seen here across Lake Arenal, it has constant Strombolian activity – frequent but moderate eruptions that draw visitors from across the globe.

OPPOSITE
Continuously glowing rock avalanches hurtle down the west flank of Mount Arenal, March 30, 2001. My fellow volcano-watcher, Isamu Kitafuji from Japan, who had observed ten earlier eruptions, described this one as quite minor.

My guide for the arduous, supposedly 6-hour trek up Mount Arenal turned up with no food or water, no pack and no hard hat, just a T-shirt and bare feet. When I questioned this lack of preparedness, he proudly produced a pair of gumboots from a woven shopping bag.

After a while, I was able to allay my fears of big-cat attack: the hot, sticky jungle had given way to a more familiar landscape of steep cinders and scoria ridges. In the fog it was hard to see where we were going, but it soon became obvious we weren't going around the mountain to the safe side. My guide had decided to show me a new, and until now untried, route to the summit – going halfway up and then across the two active pyroclastic flow paths, the very same ones I had watched in near-continuous rockfall the previous night. There was only about ten metres' visibility on the mountain, so by the time I realised where we were, we were on a small ridge between the two active flow channels – islands in the stream, should the mountain choose to send anything super-hot our way.

Large rocks travelling at 100 kilometres an hour make a roaring, whistling noise, like falling bombs in a war movie. I know, because several salvoes of these shot across the place we had just walked through. My guide took the news in pretty good heart when I told him I was making the decisions from here on, and we were going to cross the less active channel as quickly as possible to get into the bush. No sooner had we done this than tonnes of smoking boulders flew past the place we had just traversed.

Once into the sort-of-safe area, we followed a trail of debris up to the wreckage of a passenger plane that, in 2000, had slammed into the side of the volcano 200 metres from the summit, killing all ten people on board.

The bodies had been removed, but the wreck site had not otherwise been visited since the accident and was pretty much the way it had been. Large amounts of luggage and other items were scattered over a wide area, making it easy to imagine what the passengers had been doing in their last moments: taking photos and videos, reading the *Lonely Planet* guide to Costa Rica ... In the pilot's case, perhaps he had been reading the plane's instruction manual, which was lying conspicuously amid the wreckage, dog-eared, water-stained and poignant.

We got as close to the summit as we could, but reaching it was impossible, with explosions lobbing large rocks over the crater rim. It was when we started down that the fun really began. The clouds had luckily cleared, affording a fair view of the route in case there were more detours in store. In fact we got off the cone without incident – only to spend the next two hours blundering about in the hot, trackless, festering jungle before I decided I'd had enough. We had about three hours of light left, and there was no way I was sleeping in that jungle with jaguars, 20 species of deadly snake and heaven knew what other creepy-crawlies.

Resuming command again, I told the guide to follow me back to the bottom of the rockfall area. Three hours later, we came out at a small lake below a tourist observatory. It was dark, and I'd lost part of my camera tripod, along with my temper. I radioed Isamu who, according to plan, was waiting with the vehicle, and keeping the police under observation in case they got wind of my attempt on the summit. I was long overdue, and agitated. Halfway through the call, I accidentally changed the channel on the radio. The sudden loss of contact convinced Isamu I'd had an accident. He immediately contacted the police, who then resourcefully started firing their guns into the air to attract our attention. Our intention had been to escape the attention of the police altogether, but then it had also involved my guide knowing where he was going and what he was doing. When the shooting started, he crashed off into the undergrowth with no torch.

I decided the best thing to do was to walk up the track with a light on, thinking, mistakenly as it turned out, that the volley of gunfire was probably just hunters. I emerged from the track, 13 hours after setting off on my six-hour climb, to find the police waiting for me. The matter was cleared up in the normal way – with a few laughs and considerably more pesos. My guide, meanwhile, was nowhere to be seen.

The remains of a single-engine Cessna that crashed 200 metres from the crater of Mount Arenal in August 2000, only minutes after taking off from the nearby airport. Local newspaper *The Tico Times* reported that rescue workers, exhausted from trekking 4 hours to the site, had to contend with toxic volcanic gases and the fear of landslides and eruptions. Just five days earlier, a guide on the mountain had died of injuries received during an eruption.

From Costa Rica I travelled to Ecuador, where my plan was to tackle the three highest volcanic peaks, including Ecuador's tallest mountain, 6310-metre Chimborazo, which lies 1500 metres south-west of the capital, Quito. My first attempt, though, would be on the 4784-metre stratovolcano Guagua Pinchincha, another of the world's most active volcanoes. During a violent eruption in 1993, two young volcanologists who ignored a radio warning had been killed there by what is known as a phreatic eruption – a blast caused by the heating and explosive expansion of underground water. Despite my Arenal experience, I still believed in the general principle of guiding and on May 6, largely thanks to the proficiency of my guides, I completed the climb without significant incident.

Thus encouraged, on May 9 I hired three guides to assist me in an attempt on the summit of 5896-metre-high Cotopaxi. Cotopaxi has been a bad-tempered sort of hill throughout human history, with its most violent recent episodes taking place in the eighteenth century, when one killed 800 people. Since its last eruption in the 1940s, however, it had been relatively quiet. The greatest danger I would face in my attempt on the summit would be the mountain's great height.

Cotopaxi was meant to be a warm-up for the attempt on Chimborazo, but it didn't quite work out that way. We climbed without incident to the refuge at 4800 metres, and close to midnight on May 10 left for the last push to the summit, making the most of the favourable ice conditions that occur after an overnight freeze. Cotopaxi is a big, high, cold mountain, and it was my third high peak in two months. I didn't feel 100 percent well when we got underway, but I expected the fatigue and woolly head to improve as my blood got moving. Unfortunately they didn't. In fact, they got worse. I developed an evil headache and began showing the classic signs of altitude sickness. By the time the guides, concerned at my condition, called a halt we had reached 5300 metres. Although we were a mere 500 metres from the summit, they urged me to abandon the attempt and I agreed: I've always regarded getting to the summit of mountains to be optional but getting down is not. The descent, on which I was blinded by my headache and not in full control of my limbs, was hell, but thanks to my guides I made it down safely. Cotopaxi and Chimborazo remain in my in-tray.

The village of St Pierre below cloud-shrouded volcano Mount Pelée on the French Caribbean island of Martinique. St Pierre was rebuilt after being totally destroyed in 1902 by a *nuée ardente*, a huge cloud of superheated gas, ash and rock. Martinique is on the Lesser Antilles volcanic arc, which stretches 850 kilometres from Venezuela to the Virgin Islands.

A year later I was back in the region, touring some of the interesting peaks of the Caribbean. It was 100 years since the catastrophic 1902 eruption of the 1397-metre stratovolcano Mount Pelée in the French territory of Martinique, when a huge *nuée ardente* – a glowing avalanche accompanied by a billowing cloud of volcanic gases, ashes and dust – had rolled down the mountainside, destroyed the town of St Pierre, and killed 30,000 people. It was one of the deadliest eruptions ever and gave its name to peléan eruptions, the kind which happen decades apart with little significant activity in-between. Miraculously, there was a sole survivor: a man locked in a little stone prison cell was sheltered from the lava bombs and ashfall, and found by rescuers a few days later. As I surveyed his tiny refuge, I wondered what he'd done to be imprisoned, and whether he had decided to mend his ways after the experience!

The sole survivor in St Pierre of the 1902 eruption was a 25-year-old convict who was incarcerated in this tiny cell at the time. Although badly burned, he survived for four days before being rescued.

The mountain's last significant activity was in 1932, and St Pierre is again a prosperous, attractive little town – for now.

From Martinique I travelled to Montserrat, in the Leeward Isles of the West Indies, where a Pioneer Productions crew was shooting a documentary on volcanoes for Discovery Channel. The object of their interest was the town of Plymouth, which had been totally destroyed in 1995 when the 915-metre Soufrière Hills volcano erupted, sending down huge lava flows which buried everything and everyone in their path. This was one of the rare occasions when I inspected a volcanic disaster zone by the book, applying for a permit to enter the prohibited area, and even observing its conditions. I toured the ruins of the town filming and taking pictures with the Pioneer crew. It was one of the most surreal scenes I have ever seen, like something from a Steven Spielberg movie, or the aftermath of an atomic bomb.

I left at the end of the day feeling every inch the docile member of a tour group. At eleven that night, however, as I was packing my gear I found I was missing a small backpack which contained my stills camera and a mini video camera, as well as assorted odds and ends. My mother's words at times like these came back to me: 'Where did you last have it?' Suddenly I had a clear memory of setting it down in the middle of what was left of Plymouth. There was nothing for it but to sneak back into the town, walking the five kilometres from the hotel in total darkness, to retrieve it. I just can't seem to stop defying volcanic exclusion zones…

The ruins of downtown Plymouth on the island of Montserrat, photographed on June 21, 2002. About 5000 people were living in Plymouth in July 1995 when the volcano suddenly came to life after centuries of dormancy. A series of eruptions over the next two years buried settlements, made two-thirds of the island uninhabitable, and submerged Plymouth like a modern Pompeii.

Despite temperatures around 40°C (104°F), the porters (shown here) we hired for the trek up Mount Semeru brought no water of their own, leading to a dangerous incident on the mountain.

DYING FOR A DRINK

One of the strangest episodes in my career to date occurred early in October 2004, when an Australian actor and I were camped close to the top of Indonesia's Mount Semeru. Our plan was to film the volcano's activity and re-enact an earlier event for a documentary Pioneer was making for Discovery Channel. Semeru is dangerous. At 3676 metres it's the highest volcano on Eastern Java, and a stratovolcano with a recent history of violent activity. It had claimed its last victim in August 2000, when an explosive eruption had surprised a scientist at the crater's edge. It could easily have claimed another in 2004.

We hired locals to guide us and act as porters, and as so often seems to be the way they showed up without supplies. Stretching a point, food wasn't critical, but to tackle a day's hard climb in temperatures around 40°C (104°F) without copious quantities of clean water was sheer madness. As it was likely they'd try to steal ours, I took my by now standard precaution of putting tape around the necks of my water bottles and told my assistant, Richard McNair, to do the same. We also asked a park ranger who spoke English to explain to the porters that our water was not to be touched. Every now and then, as we struggled along in the jungle towards the cone, I'd look around and catch one of the porters about to crack the seal on one of my marked bottles and would put a stop to it. Mostly, though, I saw them drinking from unmarked bottles and thought nothing much of it.

At last we reached a ridge just below the hut where we were to stay for the night before going on to the summit the next day. By now both Richard and I were now knocking back water as though it were going out of style. In a sense it was. As we got closer to the hut Richard was in a bad way, running a temperature and feeling terrible. He began to have attacks of nausea and diarrhoea. The more fluid he lost, the more water he drank to try and stay hydrated. Unbeknown to us, however, the water was contaminated: the porters had drunk the water from Richard's bottles and replaced it with water from the lake, which was seething with micro-organisms.

According to a doctor in New Zealand, whom we called for advice on my satellite phone, Richard now had bacterial toxin or a virus. To keep him cool, we pitched a tent and I went to a nearby spring, collected more water and filtered it. However he couldn't keep anything down. By now he was badly dehydrated, and getting worse. I knew how dangerous this was. After a day's rest, however, Richard had recovered enough for us to walk out to the nearest sizeable town, Lumajang. Here, while still showing symptoms, he did not appear to be any longer in serious danger.

Four days later, though, his heart rhythm became irregular and we decided he needed hospital care. There was a hospital of sorts in the town, but there were neither the trained staff nor the equipment to help. Through an interpreter, I learned that the hospital owned an electrocardiograph unit. I shouted, demanded, and subtly displayed fistfuls of rupiah but was told Richard would have to wait his turn. Someone else was using the unit.

I've been in some dangerous scrapes in my time, and I've always coped. A big part of surviving in an emergency is staying calm and analysing the situation, asking yourself what is going to kill you. Once you've identified the risk, you can take steps to minimise it. And strangely, something about this process helps keep you calm. But it's different – and worse – when it's someone else in danger. And different again when it's someone you've brought into the situation and are, therefore, responsible for. Helplessness, the inability to do anything, is harder to deal with than anything else.

I got on the phone to a friend in New Zealand who is a trained paramedic. He suggested drugs and measures that could help, and when the electrocardiograph unit finally appeared told me how to set it up. As soon as we had the machine up and running, I took a trace and described it to my friend. It showed that Richard's heart was very erratic. The medics in the hospital, presented with this information, said they had no idea what was causing it. All the available medical data pointed to the cause not being dehydration. Richard's electrolyte levels were analysed and found to be within normal limits. Both the doctor and paramedic in New Zealand were baffled. The only thing that it could be, they felt, was pericarditis, a life-threatening condition. They suggested various drugs be administered, and said we should be prepared to move Richard in a hurry should his condition deteriorate. I did the only thing I could, which was to pay the owner of the only available ambulance to stand by.

Meanwhile, the doctors prescribed different drugs and said a drip wasn't necessary. Every now and then, to my intense frustration, someone would come in and remove the ECG unit for another patient, but it never seemed to be too long before it came back again. This was not entirely reassuring: perhaps the patient had died.

After a few anxious hours, many phone calls and lots of tense conversations through an interpreter, Richard seemed stable enough for us to risk putting him on the next flight back to New Zealand. We later learned that after he arrived he had not gone to hospital as we expected, but to a heart specialist who prescribed the same drugs he had been taking in the Indonesian hospital.

Shortly afterwards, I flew back to Indonesia. I hired another set of porters – ensuring every one had his own adequate personal water supplies! – and returned to Semeru. This time I made the summit and shot the film. Mission accomplished – but what a mission.

Mount Semeru, the highest mountain on the Indonesian island of Java, is seen from Lake Ranu Kumbolo, October 8, 2004. Very steep-sided, Semeru is a dangerous volcano that has erupted over 50 times in the last two centuries. Since 1967 it has been in more or less continuous eruption, and has killed over 500 people.

Cyclones, Hurricanes & Typhoons

How to Cook up a Storm

To get a proper storm going, you need the right things to happen in the right places and at the right times. To start with, you need a good-sized block of warm ocean. Scientists reckon it has to be around 50 metres deep, and a minimum of 26.5°C. This is a cyclone's fuel tank, the reservoir of energy it needs to construct itself.

Then you need the right atmospherics above that block of water. There has to be a steep temperature differential between the lower and upper layers of air, so that moist air rising will cool as it reaches great height. More than this, you need the middle layers – around 5000 metres above sea level – to contain a reasonable amount of moisture already, and you need low vertical wind-shear. Air tends to sit

Satellite photo of Hurricane Katrina making landfall on the Mississippi coast, August 29, 2005. It would turn out to be the third deadliest hurricane in United States history.
NOAA

around and move in layers; you can have different wind directions and speeds at different altitudes above the same point on the Earth's surface. Vertical wind-shear is the difference between wind speeds at different altitudes.

If any or all of this happens too close to the equator – within 500 kilometres – then you're not going to get your storm off the ground. This is because the closer to the equator you get, the weaker the Coriolis effect becomes. The Coriolis effect is the force exerted on the atmosphere by the rotation of the Earth. In the northern hemisphere, winds circulate clockwise; in the southern, they go anticlockwise. The bottom line is that you need a reasonable amount of rotation happening in order to achieve 'near gradient wind balance' – a stable area of low pressure at the centre of your storm system.

Now you're ready, but none of the above will create a storm for you. What you have are the necessary conditions for intensification to occur – a kind of high-performance workshop for hotting up existing storms. Should a suitable-candidate weather system happen along, with a modest level of organisation and low-speed, circular winds, you can get to work.

What should happen now is the heat and moisture released by all that nice warm ocean will feed into the depression, rising rapidly into the cold upper levels and generating a large thunderstorm – or, as the scientists prefer to call it, a mesoscale convective complex. Within this, a vortex may form around a core of warm air. Feed just the right amount of warm, moisture-laden air into this area, and you'll be setting up a powerful convection. Air pressure in the core will drop as the warm air rises; the water vapour will condense as it encounters the cold air around it, releasing the energy that is latent within it.

The energy thus released drives two effects: intense convection, and strengthening horizontal winds around the very low pressure of the core. If conditions remain favourable, the system will continue to intensify until, viewed from a weather satellite, it has developed into a dense, cotton-wool spiral arranged around a discernible eye. The more pronounced this organisation and the clearer the eye, the bigger, tougher and meaner the storm.

Tropical cyclones with maximum sustained surface winds (as opposed to the occasional, much higher gust) of less than 17 metres per second (34 knots, 60 kilometres or 39 miles per hour) are called 'tropical depressions'. Once these wind speeds are exceeded, the system is typically called a 'tropical storm' and assigned a name by whichever of the tropical cyclone monitoring centres happens to have jurisdiction over that area.

If the surface winds reach 33 metres per second (64 knots, 120 kilometres or 74 miles per hour) the storms are variously called:

HURRICANE the north Atlantic Ocean, the north-east Pacific Ocean east of the date-line, and the South Pacific Ocean east of 160°E

TYPHOON the north-west Pacific Ocean west of the date-line

SEVERE TROPICAL CYCLONE the south-west Pacific Ocean west of 160°E or the south-east Indian Ocean east of 90°E

SEVERE CYCLONIC STORM the north Indian Ocean

TROPICAL CYCLONE the south-west Indian Ocean

The amount of energy generated by a hurricane is mind-boggling. You can calculate it by working out how much energy is gained from the conversion of water vapour to rain (that is, from the rainfall involved in an average hurricane), or by the amount of energy a hurricane dissipates when it encounters land (that is, from the average wind speeds across the radius of the storm).

The first method yields a figure roughly equivalent to 200 times the total world capacity for electricity generation, the second a figure equivalent to half of it. The bad news is, either way, that's a lot of energy. The good news is that most – around 400 times as much – of that energy is used to generate clouds and rain than is used to drive the horizontal wind speeds.

The other bad news, of course, is that the wind speeds are plenty strong enough to do us damage. Severe tropical cyclones (as we call them in our neck of the woods) are ranked by the Saffir-Simpson Hurricane Scale.

Saffir-Simpson Hurricane Scale

	WIND STRENGTH	STORM SURGE	TYPICAL EFFECTS
Category 1	119–153 kph 74–95 mph 64–82 knots	less than 1.9 m less than 4–5 ft	Minimal. Negligible house damage. Damage to some crops, trees and caravans. Small craft may drag moorings.
Category 2	154–177 kph 96–110 mph 83–95 knots	2–2.5 m 6–8 ft	Moderate. Minor house damage. Significant damage to signs, trees and caravans. Heavy damage to some crops. Risk of power failure. Craft may break moorings.
Category 3	178–209 kph 111–130 mph 96–113 knots	2.5–4 m 9–12 ft	Extensive. Some roof and structural damage. Some caravans destroyed. Power failure likely.
Category 4	210–249 kph 131–155 mph 114–135 knots	4–5.5 m 13–18 ft	Extreme. Significant roofing loss and structural damage to buildings. Many caravans destroyed and blown away. Dangerous airborne debris. Widespread power failure.
Category 5	more than 249 kph more than 155 mph more than 135 knots	more than 5.5 m more than 18 ft	Catastrophic. Extremely dangerous with widespread destruction.

Hurricane Georges approaches the Florida Panhandle, captured in a satellite image on September 25, 1998.
NOAA

Rendezvous with the Eye – Hurricane Georges, USA

I've been fascinated by big storms since I was a kid, and when I see something brewing on one of the extreme weather sites on the internet, I still get the same tingle I used to get when the weather map showed a system with a 'T' in middle – a tropical depression – tracking southward. When a nasty bit of weather dropped south from the tropics, my dad's generation would call it a 'hurricane' – a hangover from the days of immigration from the United Kingdom and northern Europe. Today New Zealand television newsreaders and weather forecasters call such events 'cyclones'. Officially, though, they're 'storms of tropical origin'. They may have begun life as true tropical cyclones, they may have hung on to the names given them by the Fiji Tropical Cyclone Warning Center, they may even have dished out a hiding to our Pacific neighbours – usually Vanuatu or New Caledonia – but by the time they get into our colder latitudes, they've lost their oomph. We get the leftovers.

Two big storms hit New Zealand in the 1996–97 season. Cyclone Drena dropped by in December 1996 and caused a bit of havoc – boats breaking free of moorings and being smashed up on the shore, roofs lifted off, and trees and power lines brought down. Parts of the country also suffered some flooding. Then, a few months later, the even stronger Cyclone Fergus hit, causing extensive damage to the east coast. But by the standards of real storms – the true tropical cyclones, hurricanes and typhoons I'd read about since I was a kid – these hardly rated. It was a long-standing ambition of mine to catch up with a proper storm one day. There was obviously no point waiting for it to come to me: I was going to have to go after the storm.

By 1998, I'd scratched together enough money and felt the need for a bit of a holiday. The next time a big storm looked as though it were lining up a populated area, I told myself, I'd go and see if I could get among the action.

As the year wore on, though, the heat went out of the southern hemisphere cyclone season. I began paying more attention to the north, particularly the hurricane belt on the North American coastline. There were one or two false alarms. Then one morning in early September the National Hurricane Center reported a 'tropical wave' moving off the African coast. So an Atlantic storm is born. An area of intense low pressure is caused by an air mass, super-heated over the African landscape, rising and leaving a hole in the air behind it. Depending on what atmospheric conditions this hole encounters as the rotation of the Earth drags it out over the ocean, it can gain energy. As it crosses the sea, it draws the moist ocean air into its system – arming itself with the water with which it will later lash the land. If the atmospherics are unfavourable, however, the hole will fill before a cyclone can ever be generated.

This 'tropical wave' was designated 'Tropical Depression 7', and the discussion appended to the report observed that the 'SHIPS [Statistical Hurricane Intensity Prediction Scheme] model brings the system to a hurricane and so does the official forecast.'

As I monitored this storm, it began to look like the real deal. The reports mentioned 'deep convection near its centre' and a 'large envelope', and complimented the evolving storm on its 'improving organisation'. It was upgraded to a tropical storm on September 16, and within another 30 hours it was reported to have acquired 'all of the classical features to become a major hurricane' – which is to say, it had assumed that beautiful, cotton-wool, spiral shape visible in satellite photographs, curled like a snail around a distinguishable eye.

For a week, I held the stare of that single, malevolent eye on my computer monitor, daring it to blink. Meanwhile, hurricane warning facilities in the Caribbean and right along the south-eastern seaboard of North America watched it coming like a deer in the headlights. By now it had a name. Since it fell within the jurisdiction of the Hurricane Warning Center in Miami, Florida, the Center had the honour of plucking the next name from the alphabetical list it holds to christen ten years' worth of storms. Georges had been born.

Over the next few days, Georges began his destructive romp through the Caribbean. His first landfall was Hispaniola to the east of the Bahamas, but his fringes also brought misery and devastation to a number of other land masses, notably the Dominican Republic, Cuba and the Bahamas themselves. These were poor countries, with little or no preparedness for such events, despite their frequency, and with buildings and infrastructure ill-suited to withstanding the fearsome energy of storms such as Georges. Hundreds died – nearly 700 in all – and the economic damage was incalculable.

There are plenty of storm-chasers in the United States – intrepid cameramen and women, reporters and just plain thrill-seekers – who will travel the width of the North American land mass to put themselves in the path of a hurricane. But as far as I am aware, there's no one else in the world who travels the *globe* looking for a big blow. While I was sitting in my front room in Auckland, New Zealand, watching Georges growing in size and fury on my computer screen, I was almost certainly the only person outside the United States considering jumping on a plane and going after him. Most people plan their overseas holidays to avoid events such as Georges. Similarly, while there may be other people who, like me, think of a big storm as one of the greatest shows on Earth, I've yet to meet them checking into a hotel in the line of one.

The hardest part of storm-chasing is deciding whether – and when – to go. I sat at home in an agony of indecision, watching Georges wreak havoc through the Caribbean. As anyone who has watched big storms will tell you, they're tricky creatures, capable of ducking and weaving and doing the complete opposite to what they're supposed to do. But as Georges approached mainland America, the odds his path would cross a major population centre shortened. The question was becoming one not of *if*, but *when*. All that was left to work out was where he would hit, and how hard. His 'interaction with land' – as meteorologists euphemistically call the venting of a hurricane's fury and subsequent dissipation of its energy on land masses – was depleting Georges, but he was still bordering on Category 4 and had plenty of energy to lose before he became just a day of cruddy weather. With hurricanes, there's always the chance they'll do just that. The longer I could leave my decision, the surer I could be that there would be something to shoot. And the more accurately I could predict the point of impact, the more efficiently I could make my travel plans.

Then again there's an expression 'He who hesitates is lost', and that was the case here. Leave it too late, and I'd risk every airport within effective range closing, and the advance effects of the storm – flooding, high winds, to say nothing of the official response – shutting the area down. By the weekend, I'd decided it was a safe bet Georges would come ashore somewhere on the 1000-kilometre stretch of coastline between the Florida Panhandle and the Mississippi Delta. The logical place to fly to was Miami, so I booked my ticket.

During the 24-hour flight, I tried to sleep but I'm not great at sleeping on airliners at the best of times. And on this trip there was plenty to keep me awake. It was a real punt to cross the world on the off chance I could intersect the storm's path. The big adventure could turn out to be nothing more than a visit to a battered, soggy piece of the American seaside. Georges might change paths and stay out to sea, or spend his fury on some godforsaken shore, featureless and uninhabited. What a waste of money that would be. Then again, I might catch up with Georges at the peak of his powers. I had no idea what would happen then. After all, it's pretty well-known that hurricanes can be hazardous to your health.

Upon touching down at Miami, I went looking for news of Georges's latest movements. It wasn't hard to come by. In times like these, the American media really go into overdrive. The impending impact of Georges was sufficiently newsworthy to bump a raging political storm – and the 200th replay of President Bill Clinton insisting he 'Did. Not. Have. Sex. With that woman' – into second place. Not many of the networks were getting the pronunciation of the French name 'Georges' right though; they may have been coping in Louisiana, but not in Florida.

By now, early on the morning of Sunday, September 27, Georges was set to graze the Florida Keys as he made his way into the Gulf of Mexico. I had three choices: I could try to get out to the Keys to witness that landfall; I could see what happened when Georges touched down on the mainland; or I could go for the double whammy and try to catch both. In the end, a glance at a map of the route I would have to take out to the Keys was enough to convince me to stick to the mainland. There was a series of bridges to cross, and if any one of them got damaged I'd be stuck out there.

You have to wear your best poker face when you're leaning on the counter renting a car you're secretly hoping to take into floods, high winds and a maelstrom of flying debris. I got hold of a little white car of some description, and set to work driving about the Miami coast gathering shots to illustrate the preparation that goes on ahead of the onslaught of a hurricane. Georges was already having an effect on the weather, and the barrage of media reports was drawing people down to the beaches to squint at the horizons in search of the monster. Ragged clouds lit with an eerie, greenish light were marching in from the sea, and a big, angry swell was pounding the sand – classic harbingers of a hurricane's approach.

Heavy rainfall was expected, and in one place I found a group of soldiers filling sandbags and laying them up as a precaution against flooding. One, who was on exchange from the Israeli Army, was interested to hear that I was planning to chase the storm. Although he'd been through Hurricane Andrew in 1992, Sheldon Frank told me he wanted to experience a real storm. Could he tag along for the ride? This suited me: my best guess as to where Georges was going to make landfall was Gulfport, Mississippi, 1200 kilometres away. He could share the driving. And it occurred to me that the US Army identification he was carrying could come in handy at the roadblocks we were bound to strike as we neared the disaster zone.

We hit the road in short order. There was no time to lose. Although Georges had slowed down to draw in a deep breath of the warm, cyclone-energising waters of the Gulf of Mexico, he was already dumping a fair bit of rain on the coast ahead of his landfall and there was a danger that flooding, fallen trees and the usual clutter of cyclone damage would close the roads. Sheldon and I took turns driving. It was interesting to talk to him. Unlike me, he had been in a war zone: he'd done active service in the ongoing Israeli confrontation with the Palestinians.

We talked about war. He was curious to know why I would willingly put myself in the way of a hurricane but didn't fancy the perceived glamour and the thrill of war journalism. I gave him the answer I give everyone who asks this question, and I get asked it a lot. With a violent storm, an erupting volcano or a tornado, I can be fairly confident I'll be the only person on-site with a camera – or at least, the only one determined to get top-quality footage. My coverage of the event is likely to be as good as it gets, so I can afford to minimise the risks I take in shooting it. In a war situation, on the other hand, there may be 50 or 60 freelance journalists and cameramen or -women whose livelihood depends on getting the best shot. The intense competition drives them to take ever greater risks, to put themselves in more and more stupid situations in the name of that shot. The risks just become too great.

And while I haven't wasted much time analysing it, there's probably something else too. Natural phenomena such as supertyphoons and erupting volcanoes are spectacular, awesome and thrilling but utterly impersonal. War might be spectacular, awesome and thrilling as well, but there's a difference: it is the product of human nastiness. I've seen plenty of nastiness in my time, due both to nature and man. I can take it, and I can do my job and shoot it. But I don't like it, and I won't court it.

HURRICANE GEORGES	
Duration	September 15–29, 1998
Highest sustained wind speed	250 kph / 155 mph
Lowest barometric pressure	937 hPa
Damage	US$5.9 billion
Casualties	603
Areas affected	Leeward Islands, Puerto Rico, Dominican Republic, Haiti, Cuba, USA (Florida Keys, Mississippi, Alabama)

Whenever I wanted to film something, I got Sheldon to take the wheel. As we got closer to our rendezvous with the eye of Georges, I started seeing picture opportunities everywhere. Just across the state line in Mississippi we came across a five-kilometre-long convoy of big, bright-yellow linesmen's trucks, all headed in the same direction as we were. I shot roadside advertising, such as the sign at a motel near New Orleans which read 'Georges Stay Away – Everyone Else Welcome'. And I framed a road sign announcing the hurricane escape route from New Orleans in a shot of a deserted stretch of highway. New Orleans, a city of a million souls, had been completely evacuated. With extreme rainfall and high onshore winds expected to pile heavy seas on top of Georges's five-metre storm surge, the risk to the low-lying city was just too great.

As we got closer to the action, we began striking roadblocks with increasing frequency. National Guardsmen – kids with M16s slung over their shoulders – flagged us down and tried to convince us to turn around. Some looked scared, some looked excited, but plenty just looked wet and bored. But they had a job to do – and it wasn't just saving thrill-seekers from themselves. All that deserted suburbia was a looter's eldorado, and hence the towns and cities were off-limits to outsiders, and subject to a curfew for those few people who, for one reason or another, had been allowed to remain. Getting past the roadblocks was a matter of telling the person asking the questions what he wanted to hear. For the most part, American emergency workers are used to the media being around – as far as they're concerned, you know something's happened when there's suddenly a bunch of people with cameras in your neighbourhood. And for the most part, they're accommodating. Where my camera failed to convince them, we waved Sheldon's army ID. Through such stratagems, we got about pretty freely, curfews and evacuation orders notwithstanding.

All the way, the weather kept deteriorating. Gulfport, when we got there late on Tuesday, was a ghost town, complete with ghoulish voices – the horrifying shriek and mighty roar of the storm. Georges may have lost a lot of power in the last few hours before his landfall, but you wouldn't have believed it. Water on roofs was being stripped off by the wind and whirled into the night before it had a chance to pour from the eaves. Whole roofs were flapping like birds' wings, and sheets of glass in shopfronts wobbled like canvas. Crowns of trees were whipping back and forth, like head-bangers at a heavy metal gig. Pieces of steel, timber and even masonry emerged from the darkness, running and bounding like living things, and vanished again.

At around midnight, we stopped the car on a street corner where I could see some promising mayhem happening. Just how promising, we were about to find out. As I was filming, a shard of roofing iron banged into the car. I mentally apologised to the nice people at Avis and then, suddenly, the air was full of debris. Down the street, bits and pieces of wreckage were hurtling in a demented circular dance as a small tornado evolved. I pointed the camera as the remnants of buildings dilated their gyrations and began raining down on us. Behind me, Sheldon yelled. There was a loud report and a violent white flash: another piece of steel had narrowly missed us and sliced through a power line. 'Time to get the hell out,' I suggested. Sheldon wasn't arguing. As we drove, with the car bucking and rocking like a plane in turbulence, he told me he'd felt safer in his war zone. My heart was going a bit, but I'd got the incident on film. I was happy.

We found a high concrete wall and parked in the lee, breathing hard. We were pinned there by the wind – now around 200 kilometres an hour – for over four hours. At one stage I ventured into the raging night to give a report to the camera, which I'd set up in the car, pointing out of the window. Seeing I'd lived to tell the tale I was able to entice Sheldon to join me, and with the camera rolling did one of the more outrageous interviews of my career. The two of us hung on to each other and tottered about like drunks, trying desperately to make ourselves heard over the storm. Rain was pelting us like bullets. Not bad for Sheldon's 15 seconds of fame: the clip would be shown worldwide.

It was the rain – over 700 millimetres in 24 hours – and the storm-engorged sea which did the most damage. The next day we cruised about filming the aftermath. I shot the most memorable sequence, a tiled roof protruding above the surface of turbid brown floodwaters, from a small powerboat piloted by an elderly man. It was a two-storey house, he explained.

'And as you can see, it's up to my roof.'

'That's your house?' I asked him

'Uh-huh,' he said, his voice flat with shock and resignation. 'That's my house there.'

Once he'd hit the mainland Georges rapidly lost energy, and although he brought prodigious rainfall to inland areas, Gulfport was, to all intents and purposes, his last hurrah. Downgraded to 'tropical storm', he had ceased to interest the hurricane-watchers. Destructive storms such as Georges are, however, given a memorial of a kind: their names are never again used to name an extreme weather event. Memories of the devastation Georges wreaked, and his 700 victims, can rest in peace.

It was time to head back to Miami. I had plenty of time to dream up a story to account for the dents and scratches on the rental car as the 330-kilometre trip turned into a 20-hour epic. We encountered detour after detour, each manned by grim-faced emergency workers. Flooding had devastated the region and its roads. 'You sure picked the wrong time for a holiday in these parts,' one apologetic police officer told me, picking my accent. 'We just had ourselves a hurricane.'

No kidding!

Western Australia, March 21, 1999, just before the eye of Cyclone Vance hit the coast at Exmouth. Winds were measured at 267 kph, the strongest sustained wind speed ever recorded on the Australian mainland.

AUSTRALIAN BUREAU OF METEOROLOGY/ JAPAN METEOROLOGICAL AGENCY

Extreme Australia – Severe Tropical Cyclone Vance

The last year of the millennium started quietly, but I soon got to take a Pacific holiday: a week in Vanuatu trying to catch up with a Category 3 tropical cyclone. I arrived on the island of Efaté in mid January and Cyclone Dani got there shortly afterwards. Rain was the strong end of Dani's portfolio, and I got some nice shots of raging rivers and washed-out bridges, but it wasn't the real hurricane I'd been hoping for. But that's storm-chasing: 90 percent of the time you get nothing. Sometimes, of course, you have to be careful what you wish for.

Arriving back in New Zealand, I found plenty of news to film. First, a major house fire raged right round the corner from my home, and then in early February, after torrential rain, two men working on the stormwater system under the streets of downtown Auckland were surprised by rapidly rising water and swept away.

Meanwhile, I kept a weather eye on the internet. It was still cyclone season in the Pacific and a few promising events were brewing. Cyclone Elaine nearly had me packing my bags. She was blowing up nicely off the north-west coast of Australia for a while, but in the end she fizzled. However, the Timor Sea is a tropical cyclone assembly line – just ask Australians who live on the weather coast – and by the middle of March I had a map of the region on my screen again. What began as a wee low in the Timor Sea on March 16 rapidly gained intensity and began developing that pretty cotton-wool swirl that sets storm-chasers drooling. By six the following afternoon, the Darwin Tropical Cyclone Warning Centre had upgraded it to a Category 1 tropical cyclone and given it a name: Vance.

At first, Vance's track was slightly south of due west, and there was the chance he would stay out to sea, or at least vent the worst of his fury over water. However the hot, moist air there pretty much guaranteed he would gather energy rather than lose it, and Australia was a good-sized target. Sure enough, in terms Australians would appreciate, Vance bowled the Australian land mass a big in-swinger. Over the course of the 24 hours to 6 a.m., March 20, he was upgraded twice, first to Category 3 and then 4. He'd now begun to shape in towards off-stump: his track was south-west and he was threatening Australia's westernmost coast. If the trend continued, I judged, he would come ashore somewhere around Onslow or Exmouth. It was time to talk to my travel agent.

By around midday on March 20, I was on tenterhooks. The window was narrow and closing fast. As Vance approached, the likelihood of the nearest airport – Port Hedland – closing increased, and there was always the danger the outer rain bands would cause flooding and washouts. Add to this the fact the region was still cleaning up the mess Elaine had made a few days earlier, and access to the critical point was looking extremely dicey. I had to get over there. I stuffed a few things in a bag, grabbed my camera gear and got on the next flight to Perth. Once there, I switched to a flight to Port Hedland – the last one, as it turned out, before the airport closed. When I arrived in Port Hedland at one o'clock in the afternoon, I was wrung out from the flight and sweating like mad. It's hot in those parts at the best of times, and this – from anyone's point of view but my own – was hardly the best of times. The air was heavy and sticky, making for a really uncomfortable 32°C (90°F). I consoled myself with the knowledge that Vance was out there, hoovering up the moisture and the energy. This was cyclone weather.

I approached the rental car counter, trying to act casual. I was keen to rent a four-wheel-drive of some sort, but there were none to be had. The emergency services had booked them all.

'There's a cyclone on its way,' the woman told me.

'Really?' I said, raising my eyebrows.

I could have a Ford Falcon. I was vague about my plans for it, but I wasn't too fussy as we did the walk-around marking the dents and dings on the contract.

The drive from Port Hedland along the coast to Exmouth is 750 kilometres. Cyclone damage from Elaine was evident much of the way. When there are hurricanes about, this part of the country really cops it. It's a spectacular piece of coastline with vast, arid beaches and rugged headlands, but for much of the time the road plunged inland, into red scrubby nothingness. Normally fatigue would have been a factor, but I was running on pure adrenalin. I stopped at truck stops along the way and phoned my father who, back home in Christchurch, was keeping an eye on Vance's track for me.

SEVERE TROPICAL CYCLONE VANCE	
Duration	March 16–24, 1999
Highest sustained wind speed	267 kph / 166 mph
Lowest barometric pressure	937 hPa
Damage	AU$35 million
Casualties	0
Areas affected	Western Australia

Vance and I were converging on the same spot. As I got near the town of Onslow, heavy rain was starting to fall. A steady breeze was blowing. I was in the right general location, but there were still some hard decisions to be made. Should I stay in Onslow or keep going to Exmouth? Mess it up, and I may as well have stayed at home. In this game, more than any other, a miss is as good as a mile. For the storm-chasing photographer, being 100 kilometres from the action is no better than being 10,000 kilometres from it. I phoned the Perth Tropical Cyclone Warning Centre, which had taken over from Darwin and issued its first warning for the Exmouth–Onslow area two days before. They had news for me: Vance was now rated Category 5, top of the range, and he was heading my way. I phoned Dad. Exmouth was the spot, he confirmed: the ravening monster that was Vance was still bending its track southward.

'You know it's a Cat-5 now, don't you?' Dad said.

'Yep,' I replied. There was silence at the other end. He knows better than to try to talk me out of doing these things.

I arrived in Exmouth at 10 p.m. on Sunday, March 21. The town of 2500 people was deserted, and so was the hotel it seemed. At least, I had no luck in locating the proprietors. The staff of a nearby airbase were sheltering in the hotel, and I got to stay in one of their rooms. By now I was ready to drop with exhaustion, so I took the opportunity to get some sleep before the show began.

The next morning I woke at eight, and looked out to find it raining hard and blowing a gale – 50 kilometres an hour, gusting up to 100. I phoned the Cyclone Warning Centre.

'What's Vance doing?' I asked, and held my breath.

I'll never forget his reply.

'Well,' said the bloke on the other end, in that West Australian drawl, 'he's still a Cat-5 and the eye's going to come straight up the harbour.'

Right. This is it. I've got myself into the path of the most powerful weather event on Earth, what meteorologists call a 'perfect storm'. The amount of energy released by a Category 5 storm is equivalent to the amount of electricity used by the whole of the United States for same period. If Vance lives up to his publicity, he's going to be the most powerful storm to have hit the Australian mainland in recorded history. In a few hours, Exmouth is going to be the most dangerous place on Earth. And me? Well, I have a saying in situations like this: I'm either going to end up dead, or I'm going to get some awesome film footage.

There's no time to lose, so I get straight to work, touring the town in the trusty Falcon to get some 'before' shots. Exmouth used to be a bit of a backwater, a fishing town nestled halfway out on the southern headland which defines a large bay called the Gulf of Exmouth. Lately, though, the town's become trendy and West Australians have been building expensive holiday homes here. I film some of these places. A marina on the harbour is full of big-game fishing launches and gin palaces. The boats are already tossing in the wind and straining at their lines as the storm surge drives the tide insanely high. I film some of these, too. They're going to have a box seat shortly.

It's all very well plonking a huge stucco holiday home in a place like this, or mooring a million-dollar bath toy on the coast, but I wonder if their owners did any homework first? It wouldn't have taken much research to discover that this was the worst part of Australia when it comes to tropical cyclones. As in Florida, exposure to nature comes at a price. Remarkably, plenty of the buildings I film look as though they have been built with nothing more than a sea breeze and the odd heavy shower in mind.

I find the local population, or at least those who haven't had the good sense to get out of town, holed up in a hall. There's a party atmosphere, with most of the men drinking up large. I try to indicate that this might not be the ideal preparation for what's coming, but an emergency worker takes me aside and abuses me for my pains. I'm exaggerating the danger, he reckons, and causing unnecessary alarm. I shrug and go back to the hotel to get ready for Vance.

The speed at which things turn to custard surprises even me. By eleven o'clock my old mates, the banshees, are screaming as the wind picks up to around 200 kilometres an hour. Things are starting to fly apart out there. I'm being as methodical as I can, setting up the tripod in the lounge so I can film out the window, but it's not easy with the screaming and the booming and the carrying on. To keep the camera steady, I have to suspend a large container of water from the tripod as a counterweight.

By about 11.30 it's getting insane, and I'm not looking forward to doing my customary piece to camera. I have the lens aimed out the window, and the idea is to go outside and stand in the wind stream with my microphone and my little hand-held anemometer, and try to describe the sensation. This is the risky bit: if it works, it'll be great television. If it doesn't – well, I may not be around to regret it.

My first attempt is hopeless. I get more or less into position, but it's all I can do to stay upright. I open my mouth to speak, get knocked by a gust, shoot a disgruntled look at the lens, then half-stagger, half-fall out of frame. For my second try, I deploy my climbing rope and harness and, anchored in position, try to describe what it's like. The wind-driven rain is like gunfire, hammering on my head and bruising my face. The wind is roaring in my ears, with fey little gusts buffeting me about as though I'm in the slipstream of a jet. And when I say 'little gusts', don't get me wrong: these are great masses of air going faster than locomotives. Around this time, the Meteorological Office at Learmonth, 35 kilometres down the road, records the fastest sustained wind speed ever on the Australian mainland.

Soon after I've returned to the comparative safety of the hotel lounge, I realise I have to change the camera's batteries. And wouldn't you know it – the spare battery is in the car, out across the yard. Sticking a helmet on, I venture out again into the maelstrom. I get to the car okay, even get the battery out, but the wind intensifies just as I'm about to make my way back. I get down on my

stomach and hang on to the underside of the car. It is bucking and bouncing on the suspension, like a light plane in heavy turbulence. I'm just hoping it doesn't get blown over. Normally, the idea that a bit of wind could tip over a tonne-and-a-half of Australian motorcar would be a bit of a laugh; right now, it seems odds-on it's going to happen. As I'm clinging there, all kinds of junk is banging into the windward side of it. I try not to think about the people at the rental car company.

On the way back across the yard – 50 terrifying feet – the wind picks me up and hurls me across the ground to where it's been stacking everything else it's pulled up in its fury: roofing iron, fence rails and palings, palm fronds and the odd fallen palm tree. I'm giggling like a maniac, bruised and scratched, but otherwise unharmed.

The fun's not over. Even when I think I'm out of the worst of it, Vance has another crack. I'm standing on the restaurant verandah, filming water trying to run off the spouting and instead bending upward and disappearing over the roof line. Suddenly there's a groan and I feel a huge blow to my head and arms. Next thing I know I'm on the deck, tangled up in a mess of iron and timber that used to be the verandah roof. After a bit of panic, I find my camera's all right. After a bit more, I find I am too, apart from another bruise or two to add to my collection.

With Cyclone Vance breaking all Australian wind-speed records, my first attempt to film a piece for television news is a dismal failure. For my second (seen here), I tie myself into position and manage to get a few seconds of footage as trees crash down around me.

By one o'clock, as Vance's eye passed over the southern end of the Exmouth Gulf, the wind was definitely slackening. Another half-hour later, it looked as though I was going to get away with it. Most of me was bathed in relief. Part of me was exhilarated: if I hadn't got killed, it must mean I had got the footage. And a fair-sized chunk was deflated. That, in storm-chasing terms, was the Holy Grail, the pinnacle of one's career. There are plenty of storm-chasers in the world, but only a tiny fraction would ever experience what I had just lived through. I'd been as close as you could hope to get to the centre of a Cat-5 storm, the ultimate grand-daddy of them all. From now on anything less was going to seem pretty ho-hum.

The job wasn't over, of course. I had the 'before' and 'during' footage in the can. Now I needed the 'after'. By late afternoon, the wind had died away to practically nothing. Vance was now rampaging further south, leaving the ashen-faced residents of Exmouth to emerge from their refuges and contemplate the damage. What they found was as close to a war zone as they were ever likely to get. Most of their town had been destroyed, with hundreds of houses reduced to matchwood. In one place, I filmed a house whose living-room wall had been peeled off like the lid of a sardine can. In another, I shot a workstation amid the debris, complete with computer and textbook.

At the tourist park many of the caravans had been chained to their concrete pads, but this hadn't saved them. In most cases, all that was left were the wheels and chassis – still chained down but picked completely clean of their upper works. Other caravans were completely gone. I filmed people wandering about in the debris, lifting things up, dropping them again. They looked stunned. The town's trees appeared to have been mown by a giant weed whacker. Telephone poles made of ten-millimetre steel had been bent over. And in the bay the marina had been reduced by the wind and four-metre storm surge to a heap of flotsam and a chewed-up beach front. You'd never have guessed some of the debris belonged to boats if you hadn't picked out the odd bit of recognisable chandlery. I filmed a marine toilet in a mess of planking and shredded fibreglass. It seemed an apt symbol.

The havoc that Vance wreaked was, thankfully, all economic. For a time, a man was missing, believed swept away in a swollen stream, but he showed up when it had all blown over. No one died, in contrast with the awful toll taken by Cyclone Tracy, which devastated Darwin on Christmas Eve 1974, killing 65. But the economic toll was steep enough, and all the more deplorable because it could have been minimised with a bit of foresight. Cyclones are unpredictable, true. But you can say with certainty that some areas will experience them, and the coastline from Port Hedland to Shark Bay is one such area. And as climate change begins to make itself felt, there's every reason to believe cyclones will become more frequent and more powerful.

While waiting for contact with the outside world to be re-established, I helped clean up. All power, water, electricity and phones were out. Dinner was cold baked beans, and like all good storm-chasers I was carrying a water-purifying filter and could draw drinking water from the hotel swimming pool.

Exmouth residents emerged after three hours of Cyclone Vance to a scene that resembled a war zone. Some buildings had been reduced to matchwood and others, as here, thrown around, coming to rest at crazy angles. Vehicles had been overturned and boats wrecked.

The next day helicopters appeared, bringing the media. I heard a plane was planning to fly in as soon as a runway could be cleared. This would be a good way of getting my footage out to the outside world. I rounded up a few locals and a truck, and in short order we had cleared and marked 400 metres of fairly serviceable runway. After waiting for an hour or so, I phoned the pilot – no mean feat, as I had to stand on tiptoe on the roof of the car to get cellphone reception, just like the phone company ads. Once I got hold of the pilot, I learned he had decided to put down at a military strip at Learmonth. I told him I'd just spent a couple of back-breaking hours clearing a nice 400-metre strip at Exmouth. He replied he'd have needed 900 metres, anyway.

Once I'd made arrangements to get my footage out another way – through the mobile uplink a local news crew had brought in – it was time to head north again for Port Hedland. Vance had flooded most of the roads. I can report that a Ford Falcon will go reasonably well with water up to the windows. Any deeper, and you'll need a big truck with a chain to tow you through. And if you're considering buying a used rental Falcon in Australia, don't, whatever you do, buy the one I had. You'll know it: it has scratches and dents all over it, and is missing a wing-mirror. I don't know what the folks at the airport were so upset about. I got their car back to them, didn't I?

A Very Close Shave – Hurricane Kenna, Mexico

Such was the force of Hurricane Kenna in Villa Hidalgo, Mexico, that many solid buildings collapsed completely. In other affected areas, four people were killed, and in the town of San Blas a few kilometres west, up to 90 percent of homes were damaged or destroyed.

The alarm went off on the morning of October 24, 2002. I got up, yawned, stretched and went downstairs to make a cup of tea and have a bowl of yoghurt and muesli. It was shaping up to be a lovely day, and I thought I might head out to Auckland's west coast for a run along one of the beautiful black-sand beaches with which the city is blessed. While chewing on my muesli, I had a quick squiz at the internet. Half a world away, off the Pacific coast of Mexico, Hurricane Kenna was meandering north as it had been for several days. I'd been watching it without a great deal of interest. It didn't impress me as anything that would spin up to much. That morning, its peak wind speeds were nudging only 80 knots – 150 kilometres – per hour, and like most hurricanes arising in this area it was expected to tail away into the open ocean.

The only thing worthy of my attention on that lovely spring morning was that Kenna was now expected by weather authorities to make a rare about-turn and head for land, possibly even making landfall. But as far as I could tell, it still wasn't worth going halfway around the world for. The deadline for my decision was looming. I was inclined to stay. I'd been in email contact with American storm-chasers Jim Leonard and Jim Edds, and nothing in our discussion to date had given me cause to reconsider.

I wandered out to the car, threw in my running gear, and was in the process of locking my front door when I heard the little chime which signifies the arrival of an email in my inbox. I'll have a quick look, I thought, just in case it's important. It was. Jim Leonard was writing in a state of high excitement to say Kenna was now rapidly intensifying and could be packing winds as strong as 120 knots – 220 kilometres per hour – when it made landfall.

'If you want to catch a real storm,' Jim wrote, 'you'd better get your ass over to Mexico.'

Three hours later I was boarding a plane for Los Angeles. At LA International Airport, I had only enough time to put through a quick call to Jim on my cell phone before making a mad dash for the connecting flight to Mazatlan in southern Mexico. I arrived in Mazatlan at 6 o'clock at night and phoned Jim again. He reported that Kenna was still on track for a landfall 280 kilometres south of Mazatlan. The estimated time was 5 p.m. the following day – by one of those quirks of international travel, only a calendar day after I'd left Auckland. Kenna had now been cranked up the Saffir-Simpson hurricane scale to a powerful Category 4.

This was my third storm-chasing venture in this part of the world. The first hadn't gone well. In September 2000 I'd spent a week in Puerto Rico tracking the progress of a Hurricane Joyce, annoying airline staff by making and cancelling flights to just about every possible Caribbean destination as the storm meandered along, lining up first one island and then another. At the end of the week, Joyce had fizzled into just a tropical storm.

A couple of days later, however, it had become apparent that another storm, this one called Keith, was intensifying and heading towards the Yucatan Peninsula. I flew to Miami, and then on to Cancun on the first available morning flight. At the car rental office I tried to act casual, but staff became suspicious when I insisted on a four-wheel-drive vehicle. To their credit they hired me one, although they were careful to make sure I signed the insurance waivers and understood all about the excess.

By this time, 8.30 a.m. on October 1, Keith had spun up into a Cat-4 hurricane, packing winds of 210 kilometres per hour, gusting to 249. Hurricane warnings were being posted all along the coast from Belize northward.

Communications were a problem as there was nowhere I could get a link to the internet up and running. Eventually in Tulum, about 130 kilometres south of Cancun, I found an internet cafe. It was filled with tourists looking anxiously for information on Keith. I thought grimly that I was probably the only one keen to head *towards* the storm. However, I was proved wrong when I got talking to one of the people there, an American on holiday from Hawai'i. His eyes went sparkly when he learned what I was planning to do. He decided to come with me, leaving behind his less adventurous brother and girlfriend.

We drove 250 kilometres south to Chetumal, a city on the border with Belize. By this time, Keith had virtually stalled just off the coast of Belize, and he stayed this way for the best part of a day. The entire time he was stationary, his eyewall – the part just outside the eye, where the most damaging winds and intense rainfall occur – hammered the town of San Pedro on Ambergris Cay, a tiny island off the coast.

I was intensely frustrated by dramatic ham radio reports coming from San Pedro, describing 200-kilometre-an-hour winds ripping houses to pieces. Chetumal was as close as I could get to the action. The border into Belize was closed and no one was being allowed in, quite apart from the fact that all the roads were flooded and likely to be impassable. By October 3, Keith had weakened to a tropical depression and started to move across the Yucatan Peninsula toward Escarsega. As I drove in his wake, I saw people who had already been living in poverty before the storm rushing to save themselves, their few possessions and their animals from the flooding. I read reports in newspapers describing rainfall of 27 inches – 700 millimetres – before Keith moved out over the Gulf of Mexico and re-intensified, making landfall just north of Tampico, Mexico as a strong Cat-1 hurricane.

A second chance? Well no, as it happened. I couldn't get to Tampico either, because all the airports were either closed or too far afield for me to stand much chance of getting to the area in time. While I was monitoring Keith's progress, I heard a report of a plane with 60 people aboard having crashed in Mexico in heavy rain, with six fatalities. I parted company with my slightly disillusioned novice storm-chasing buddy, and consoled myself by visiting other sites on the Yucatan Peninsula, including the ten-centuries old Mayan ruin of Chichén Itzá. Those Mayans knew how to build hurricane-proof buildings.

When I had flown back to Miami at the end of this frustrating escapade, the plane had landed in the middle of a violent thunderstorm. We had been fortunate to get to ground at all: ours was the last plane allowed to land. The price we paid was to be kept waiting on the aircraft for another two and a half hours while it sat only 30 feet from the terminal. Florida's aviation rules, the captain explained, prohibited ground staff from working during a thunderstorm.

In Mazatlan, after talking to Jim I decided that instead of staying in the town overnight I would drive south immediately. This would cover me in case Hurricane Kenna sped up, or turned further south. The decision meant breaking my rule of never driving in rural Mexico at night. On television you hear about the danger from *banditos* who prowl the countryside in darkness, preying on unwary gringos. I've yet to see one. The real danger is the antics of the long-distance truck drivers. These men drive huge dilapidated rigs, often hauling several overloaded trailer units, at breakneck speed, no matter what the conditions prevailing inside or outside the cab. Rain, hail or shine, they're way over the speed limit. And whether it's thanks to tequila, other mind-altering substances or just questionable judgement, they perform the most amazing overtaking manoeuvres absolutely everywhere – on straights and blind corners, with the road ahead empty or crowded with oncoming traffic. That night I saw hundreds of instances of certifiably insane driving.

HURRICANE KENNA	
Duration	October 22–26, 2002
Highest sustained wind speed	305 kph / 190 mph
Lowest barometric pressure	913 hPa
Damage	US$80 million
Casualties	4
Areas affected	Mexico

All along the way, I phoned Jim for regular updates. Since I was travelling a north–south route, getting in Kenna's path was as simple as working out at which latitude he would intersect the coast. Jim figured that was going to be 21.5° north. All I had to do was drive south until I was half a degree south of the 22nd parallel. That turned out to be the Ruiz truck stop on State Highway 15. I found a small hotel with friendly locals who understood my bad Spanish, and whom I managed to persuade to evacuate their families. Kenna was now a dangerous Category 5 hurricane, with winds gusting to over 300 kilometres an hour. I was able to tell them with complete confidence that it was probably going to flatten everything.

I slept like a baby. When I woke at six the next morning, I fired up my satellite phone, dialled Jim and logged on to the internet, all at the same time. I got a shock when I saw the screen, so much so that I hardly heard Jim telling me that Kenna, still a Cat-5, was now going to make landfall at 11 a.m. instead of 5 p.m. His path had altered slightly, so there was no time to lose. I hit the road, and using a combination of GPS – Global Positioning System – and Jim's telephone updates I lined up Villa Hidalgo, as surely as Kenna himself was doing from the west. Villa Hidalgo is a little farming town near the sea, 30 kilometres south of Ruiz and at exactly 21.4° north. I arrived, with a couple of hours to spare, at the point where the eye was going to come ashore and the winds would be at their most destructive.

The sight that greeted me as I rolled into Villa Hidalgo was amazing. Life was going on as normal, with everyone seemingly oblivious to the impending destruction of their town. By this time I was so sure of both the storm's path and its power that I felt I had to tell the local police and army what was going to happen. It turned out they knew a hurricane was going to hit somewhere nearby. They didn't realise it was expected to be the strongest to reach Mexico in nearly 60 years, and had no idea they were in for a direct hit. Even when I explained the situation carefully in my best pidgin Spanish, I could tell they only half-grasped the enormity of it. I then showed them video footage on my laptop computer of what a Cat-5 looked like. This was more persuasive. They commandeered a fruit vendor's truck equipped with a rooftop public-address system, and toured the streets, trying to warn people to batten down the hatches and stay indoors.

An hour later Kenna arrived, bringing torrential rain and high winds. Soon I was the only one left on the street, the soldiers and policemen having fled the area in a dump truck. Cue wind, airborne debris, mayhem. *Showtime!* I tried to get spectacular footage of flying debris from under a solid concrete balcony, as sheltered as possible from the wind. Nothing much was happening, and I'd just turned off the camera with a view to relocating when all of a sudden hell broke loose. The winds increased tremendously, as though whoever was driving the storm had changed gear and jammed his foot to the boards. Debris of every description started to fall, clattering and bouncing in the road. No sooner had I started rolling again than part of a roof hit a passing car. More tiles boomed off the roof of my rental car. I swore a bit, left the camera, and drove the car up on to the footpath as hard as I could manage against the building and in its lee. Things were rapidly becoming life-threatening.

I moved the camera and tripod hard up against the wall, put on my climber's helmet and rolled down the car windows so I could get in and leave without having to venture too far from the shelter of the wall. What happened next was amazing. A steady stream of pieces of masonry and assorted other debris began to fall. Then, without warning, the stream became a torrent. A cascade of huge slabs of cement and concrete blocks blotted out my view. Everything went dark. When the dust cleared I looked myself over and found that, aside from being covered in dust and debris, I didn't have a mark on me. Better still, the camera was unscathed and still rolling. I expressed my relief at this near-miss in a short stream of earthy Anglo-Saxon, and then made a run for the car before my luck really ran out.

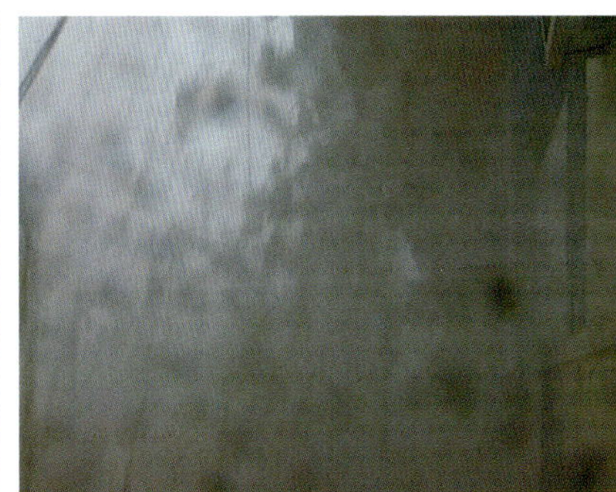

As I stood under a concrete balcony filming Hurricane Kenna hit the small Mexican town of Villa Hidalgo, the entire top storey of the three-storey building collapsed on to the street, producing this astonishing sequence of photos. Only minutes earlier I had moved my car a few yards away. By some miracle I survived.

The whole top storey of a three-storey building had come down on to the street. Glancing up nervously, I saw there was plenty more where that had come from. As anxious as I was to get out of there, it wasn't easy. A set of power lines had come down, too, and had snarled the front end of my car, which was jammed up against – and partly covered by – a pile of rubble. The cables were too strong to break. I slammed the car backwards and forwards to loosen them, and then reversed off them. The effect was to remove the car and me from the point where the building had collapsed, into an open area under threat from other buildings. With the wind approaching 300 kilometres per hour, I reversed blindly up the street. The air was choked with flying debris, continuously booming and banging into the car.

Eventually the square bulk of a big truck, abandoned in the middle of the street, loomed out of the dim, whirling void. It promised to provide good cover. I parked behind it, facing downwind. Soon, though, I felt a nudge from behind. The truck was being moved by the wind and shunting me. I had to keep my foot on the brake to keep the car – and the truck – stationary. Not far away, an empty car moved off down the street, gathered speed and hit a building. It was too dangerous to even think of getting out of the car to measure the wind speed. To do so would be like standing up in the middle of ground over which a machine-gun was firing, and about as survivable.

The storm stayed at the peak of its powers for an hour. Halfway through I rang Jim on the satphone.

'The storm's just blown a concrete building over on top of me,' I told him. 'I hope it's not getting much stronger.'

'The worst has passed,' he replied.

I watched roofs and walls passing by at motorway speeds.

'I'll take your word for it,' I said.

When it was over, several people were dead – killed, as I had so nearly been, by falling masonry. The whole town was wrecked. Every street was blocked, there were no power or phones, and the streets were littered with dead dogs and other animals. The town's major industry, a Coca-Cola factory, had completely disappeared; only the front gates remained. Every truck on the highway had been overturned or blown off the road.

When I returned to the Ruiz Hotel to collect the rest of my gear, I found my room flooded. It wasn't the first time I had had cause to be grateful I'd taken my usual precaution of sealing up excess gear in watertight, protective Pelican cases and stowing it high up. The bed was still dry, but the room was a complete shambles and reeked of wet carpet. I wandered about the deserted hotel. Eventually, I came across the housekeeper, who had bravely stayed on. She told me to help myself to a better room.

The next day I returned to Mazatlan and returned the slightly worse-for-wear car to the rental company.

'Aah … a building fell on it,' I told the clerk.

'You are not hurt, señor?' he asked solicitously.

'No, but the car's a mess.'

'Ah, *de nada*. You are insured.'

Waiting at the airport, I checked the internet. Mount Etna was shaping up for a major eruption. One day at the office ends. Another begins.

Stunned residents of Villa Hidalgo emerge to survey the devastation wrought by Hurricane Kenna, October 25, 2002. An hour earlier they had been carrying on their normal lives, completely unaware their town was in for a direct hit. Only two hurricanes, in 1959 and 1976, had been more powerful when hitting land.

A SHOCKING SITUATION

By early 2003 I'd got storm-chasing down to a fine art. Dawn on March 14 found me in a hired Land Rover on the outskirts of the town of Koné, up the coast from Noumea, the capital of New Caledonia. There with me was Category 5 Cyclone Erica – and I mean *right* there. She had arrived at 3 a.m., five hours ahead of schedule, tearing into the town with winds of over 200 kilometres per hour and torrential rain. By 6 a.m. the popping sensation in my ears heralded the arrival of the storm's centre, with a drop in pressure to a seriously low 910 hectopascals.

Using my Iridium satellite phone I contacted fellow storm-chasers Jim Leonard and Jim Edds in the United States, who were monitoring Erica's progress on the internet. At my request, they texted me regular fixes of Erica's position. Using this information and consulting my portable GPS set, I was able to do something extraordinary: for nearly two hours I drove south, actually in the eye of Cyclone Erica, as she tracked south from Koné all the way to the town of Poya down the coast. Erica brought down thousands of trees and power lines ahead of me, and laid on flash floods, but tucked into the eye I was in eerily calm, fine weather.

All was going well and the trusty Land Rover was negotiating all obstacles until suddenly, just as I noticed out of the corner of my eye that a power pole beside the road was starting to topple, the vehicle began to lose power. As the pole crashed to the ground behind me, the vehicle ground to a halt, stalled and then, contrary to every experience I'd ever had with mechanical failure, it shot backwards.

A quick check in my mirrors revealed the problem: I had snagged a set of fallen power lines, and pulled down the pole. The cables, too strong to break, had absorbed the Land Rover's forward momentum and then released the vehicle again like a bungy cord. The poles, which looked as though they had been pretty tall, were fitted with big, solid insulators. Great. These were clearly high-tension lines. I was now stuck in the middle of nowhere, in the eye of an intense cyclone, surrounded by potentially live high-voltage lines. Given the state of the lines and the absence of sparks, I thought there was a good chance they were dead. But I knew from a friend who worked as a linesman that after an outage high-voltage pulses were often sent down the wire to try and re-establish the connection. I was not keen to take a chance that they weren't live.

I opened my door gingerly and leaned out to see what kind of attachment existed between the wires and the vehicle. I wasn't about to step out, in case the bodywork was live. The cables were caught up on the front of the Land Rover, about a metre off the ground. It looked as though I could get free if I could force them down on to the tarmac and drive over them. Rummaging around, I devised a solution. I had two plastic Pelican cases of heavy equipment. I figured that if I could throw these on to the lines on either side of the vehicle that should do the trick. The trouble was, I would then have to retrieve the cases. What if I tied a climbing rope to the handles? The nylon wouldn't conduct unless it was very wet – and anyway, I could tie the rope to the vehicle and tow the cases off. At no time would I have to step to the ground until I was safely clear of the lines.

That was the plan, and it worked perfectly. Who knows whether I'd be telling the story if it hadn't?

For two hours I drove south in the eye of Cyclone Erica, a Category 5 tropical cyclone, until being foiled by a power pole that crashed to the ground, trapping my vehicle. The eye of a cyclone is literally the calm in the storm, characterised by light winds, and sometimes even clear sky above. The eye is circular, typically about 35 kilometres (22 miles) across, and surrounded by a dense ring of cloud known as the eyewall. It is in the wall that winds are at their most ferocious.

An aerial view of Tikopia as we fly over to assess damage from Cyclone Zoë, January 1, 2003. The island is the remnant of an extinct volcano; two peaks surround the ancient crater, which is now a lagoon. To our astonishment, while the island's water system and crops were devastated, there had been no fatalities. Tikopians, well-used to cyclones, had fled to mountain caves before the giant waves and winds hit.

Forgotten Island – Cyclone Zoë, Solomon Islands

It was Christmas Eve 2002 and I was on an Air Vanuatu airliner, bound for Port Vila. Sharing the flight with me were a planeload of holiday-makers looking forward to lying on white-sand beaches, snorkelling over coral reefs, soaking up tropical sun, and leaning on the bar of a fancy resort, admiring the sunset and sucking on a cocktail – all at off-peak rates. Well, there are a few good reasons the western Pacific has a tourist off-season each year, and even as we adjusted our seat-backs or listlessly fiddled with the dials on our armrests trying to find the soundtrack for the in-flight movie, one of those reasons was bearing down on our destination.

I'd had a quiet November filming the usual news stories, and so far December had been pretty boring too. Then one of the swirls in the upper atmosphere over the warm tracts of the Pacific Ocean north of Fiji began to show signs of spinning up into something interesting. By late December I was reading that 'animated enhanced infrared and multispectral satellite imagery indicates cyclic but organised deep convection with increased poleward outflow'. In layperson's terms, something was brewing. It looked as though there was going to be an adventure under my Christmas tree after all.

Cyclone Zoë – aka TC06P, the sixth tropical cyclone to be identified that season by the Fiji Tropical Cyclone Warning Centre – had been born near Tuvalu, around 340 nautical miles north of the main islands of Fiji. She had scribed a bit of an arc to line up Vanuatu, and had then begun a slow drift – less than ten knots – heading west of south-west, straight for the northern part of the Vanuatu archipelago. Weather features near the coast of Australia now looked as though they might pull her toward the south, and developing westerlies threatened to draw her sting within a few days. In the meantime, though, she was gathering strength and looked certain to make landfall in the Vanuatu group, possibly near Port Vila itself, or on Espiritu Santo, a large island in the northern group.

By the time I was on the ground in Vila, Zoë had ratcheted sharply up the intensity scale. She was now packing 200-kilometre-an-hour winds and 15-metre swells. And while her progress was slow, it was steady and unwavering, curving nicely toward the south. I booked into a hotel and watched and waited.

By December 27 it looked as though Zoë had vindicated my decision to make the trip: she had become a monster, a Category 5 storm shaping up to be the most powerful ever to hit the region. The only problem I had was that she had slightly altered course, flattening her trajectory toward the west, and was now looking to either paste the Banks Islands, the most northerly of the Vanuatu group, or miss land altogether. I got myself up to Santo on a scheduled Vanair flight, only to watch in dismay as Zoë's track bent north again. She was running out of targets. To the north of the Banks Islands, there was a wide tract of open ocean between Vanuatu and the southernmost of the Santa Cruz group in the Solomon Islands. It was odds-on that Zoë was going to thread through the gap.

CYCLONE ZOË	
Duration	December 23, 2002 – January 1, 2003
Highest sustained wind speed	240 kph / 149 mph
Lowest barometric pressure	879 hPa
Damage	unknown
Casualties	0
Areas affected	Solomon Islands

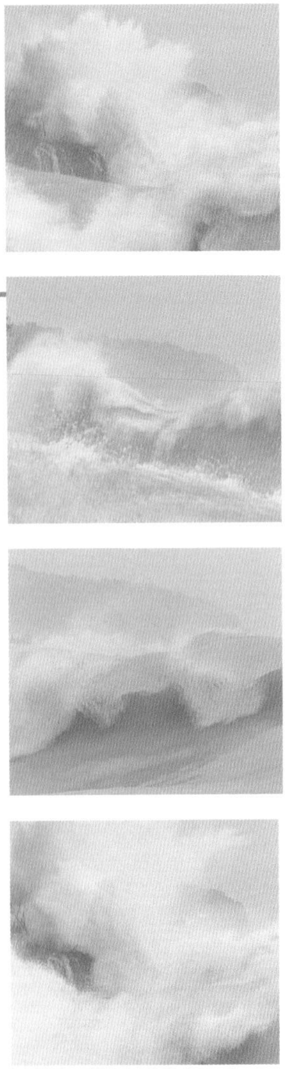

Huge waves break at Hog Harbour on the island of Espiritu Santo, Vanuatu, as Tropical Cyclone Zoë passes. These photographs are taken from video film.

My fears seemed to have been realised the following day, when Zoë missed Vanuatu altogether and slowed up to the south-east of the Santa Cruz islands. When I looked at the internet, her image – one of the prettiest satellite photographs I had seen, with a deep, crisply defined eye and a cotton-candy swirl of rain and thundercloud – taunted me.

I mucked about on Espiritu Santo for another day. All the meteorological signs were that Zoë would be drawn south in the near future, but as Sunday, December 29 dawned and dragged on it became clear I was wasting my time hoping for a significant hit anywhere in Vanuatu. There didn't seem to be anything to do but fly back to Vila. Once there I made preparations to head home to New Zealand. At the hotel, I logged on to the internet to look at the latest information. Sure enough, Zoë was still out there in the middle of the ocean, having covered only 100 kilometres or so, all pretty much due west. From a chaser's point of view, all that tropical cyclone goodness was going to waste, vented on the sea and a few chunks of rock.

While I was studying a map of the area, one of the bits of rock caught my eye. It was called Tikopia. It was tiny – some maps didn't even deign to show it – but clearly someone had cared about it enough to give it a name. Perhaps, I thought, there were people who called it home. I did a Google search on 'Tikopia', and came up with a handful of hits, mostly anthropological sites. I learned that Tikopia was an atoll, the remnant of an ancient volcanic crater which these days consisted of two low peaks bracketing a lagoon and surrounded by a fringe reef. It was about 15 square kilometres in area, and much of it was rugged and uninhabitable. But the habitable margins *were* peopled. There were around 1500 souls living there, currently enduring the full fury of Cyclone Zoë.

This had the makings of an interesting story.

The internet couldn't help me with the other burning question: How to get to Tikopia? The island had no airstrip, and only an occasional boat called there. That was no good to me. A helicopter would be ideal, but there seemed to be only one in Vanuatu, and it wasn't cheap to charter. I decided to go back to Espiritu Santo and see what I could arrange from there. So on the last day of 2002, for the third time in three days, I found myself on a little aircraft cruising over the beautiful, rugged interior of Malakula, Vanuatu's second largest island. Out to the west, I could see a smudge of smoke where sulphurous Mount Marum smouldered away on the volcanic island of Ambrym. To the north, there was a great pall of cloud. The further north you looked, the dirtier the cloud. The outer bands of Zoë had cruelly hooked north-west towards Tikopia and parked there, lashing the tiny island with the full fury of her eyewall winds and the monstrous swells of her storm surge.

As we landed at the aerodrome at Luganville on Espiritu Santo, I couldn't help noticing an orange and white Cessna 182 parked by the strip. I asked a few locals in the little shed which passed for a terminal, but no one seemed to know anything about the plane – whose it was, or whether it was for hire. After some quick thinking, I asked to be directed to what I expected to be the focus of expatriate culture on Santo and the hub of local commerce: the nearest hotel, and more specifically its bar. My instinct proved correct. There were quite a few people in the bar of Hotel Santo, and while no one was wearing a flying helmet I fancied my chances of finding out about the plane.

I cleared my throat. 'Excuse me,' I shouted. 'Does anyone know who owns the Cessna parked at the airport?'

After a moment, a stocky man with a large, greying, ginger moustache and – good sign – aviator glasses detached himself from the throng of patrons.

'Yeah, that's my plane,' Harold Tavner said, in an unmistakably Australian accent.

I bought him a beer and set about telling him all about me, Cyclone Zoë, Tikopia and what I had in mind for his plane. Harold didn't reject out of hand the notion of flying up there. He didn't know where Tikopia was, so we studied the marine chart I had with me. He measured it out roughly. Tikopia was around 160 kilometres north-west of the Banks Islands, which were roughly 140 kilometres north of Luganville. Harold happened to know that there was an airstrip on Mota Lava in the Banks Group: he was the man who cut the grass on the 40 or so strips operated by the Vanuatu Aviation Authority, including that one. We could, he reckoned, fly to Mota Lava, refuel, and comfortably make the 320-kilometre round trip to Tikopia and back. He routinely flew longer distances within the Vanuatu group.

The only problem was navigation: his plane didn't have the necessary instrumentation for a long ocean crossing, nor did he have any experience of flying out of sight of land. I told him I had a hand-held GPS set. We could regularly plot our position on the chart using that and thus find our way. How hard could it be? After a few minutes' thought, Harold declared himself willing to have a crack at it, provided I paid for the fuel. He wasn't supposed to use his plane for commercial purposes, so he couldn't accept a charter deal. The thrill would be enough, he reckoned. I quietly reckoned he might be right.

Pilot Harold Tavner refuels his Cessna 182 on Mota Lava airstrip, Vanuatu, on our way out to reconnoitre the damage to tiny Tikopia from Zoë, a Category 5 tropical cyclone.

Early on New Year's Day, Harold turned the key in the cockpit of his little plane. The engine whined over and roared into life. We bumped over the grass to the end of the runway, accelerated along it and were airborne. Within minutes the thatched roofs, coconut palms and taro plantations, slashed by a muddy road, had given way to mottled azure and finally Prussian blue ocean. Out to the west, you could see the seriously dirty weather which represented the remnants of Zoë, now downgraded to Category 3 but still a powerful storm. After battering Tikopia for the weekend, Zoë had abruptly headed south-east, as though she had something urgent to attend to in Fiji.

As we flew I consulted my portable GPS, now perched on the plane's dashboard, to get fixes on our position, and plotted them on the chart. This wasn't yet necessary: we could see Mount Garat, the highest point of Santa Maria Island, ahead of us, and beyond it, blue in the distance, the elevated interior of our immediate destination, Mota Lava. But we needed to know before we lost sight of land whether our crude navigational system was going to work. I had no wish to do an Amelia Earhart and run out of fuel over open ocean.

All systems seemed 'go' when we landed on the grass at Mota Lava. My plots and the course we had actually taken seemed to tally up. While the plane was refuelled from a drum of aviation gas, and after Harold had cast a critical eye over the length of the grass on the runway, we used the chart and a protractor to work out a course for Tikopia. We were both reasonably confident, but then noticed something amiss. 'Oh, shit,' Harold said mildly. 'We had the protractor upside down.' That was easily fixed: we would just steer a reciprocal course. Soon Vot Tande, an extinct volcano and the last of the Banks Islands, had dropped away behind us and become obscured by cloud. Now we really were relying on chart work and the GPS. You feel pretty small in a Cessna when all you can see out the window is the vastness of the Pacific Ocean.

Harold was curious to know what we would find when we got to our destination. I told him about Category 5 storms, the power they pack and the damage they can do. He had seen some big blows in his 15 years on Vanuatu, but few people had ever experienced a storm like this. The construction of the dwellings on a tiny place such as Tikopia was likely to be primitive and completely unable to withstand the destructive winds, and most of the habitation would have been on the coastal margins, exposed to the storm surge. The islanders' crops, water supply and almost everything else would have been destroyed. If there hadn't been large-scale loss of life, it would be a miracle. I warned Harold that we should prepare ourselves to see total devastation, with bodies everywhere.

According to the GPS, we were now getting pretty close. We flew out of a cloud and there, right in front of us, were the twin peaks of Tikopia. Harold dropped us down, and we flew low over the island. Sure enough, everything was flattened. It looked as though an atomic bomb had been dropped. Just about everything green had been stripped off the vegetation, much of which was smeared across the ground. You could see where villages had been: timbers protruded from the sand. And the waves had deposited chunks of coral right up to the foot of the peaks, along with wrecked boats and canoes.

Tikopia, showing the remains of a house flattened by the cyclonic winds. Nearly all the dwellings on the island had been destroyed, as had most of the vegetation. Amazingly, no one had been killed.

Then we began to see bodies, live ones, people running out of what was left of the bush and waving to us. Many were holding up sheets of white plastic to attract our attention: others were just standing on the beach shading their eyes. While I took photographs, Harold did a rough count and got up to a hundred people. We didn't see any corpses at all.

There wasn't a lot we could do for the people down below, but we hoped our appearance would reassure them they hadn't been forgotten. We began to wish we'd brought some kind of aid supplies. Anxious to preserve our margins of safety, Harold turned us back for Mota Lava. On the way we talked about what practical help we could have given the Tikopians. Both of us thought a radio would have been of most use. A two-way set could have been padded up well enough to protect it from the impact, and we could have pitched it out the window as we flew along the beach. At least that way the islanders could have established contact with the outside world. Next time, I reckoned, that's what I'd do.

We were both excited, not only to have made it to the island safely but more importantly to have seen signs that significant numbers of Tikopians had survived. As soon as we were back on the ground, I knelt on the airstrip and set up my satellite phone to send the pictures I had taken – along with a brief written report – to TV3, via Doug Keisling, my agent in America. Then we flew back to Luganville and headed to the bar at the Hotel Santo where it all began, for what we considered to be a well-earned beer.

Zoë was a fair-sized storm, but she was nothing beside the one I stirred up with that story. My pictures screened around the world, making headlines on the BBC, CNN and Australia's ABC. They were good pictures, but what captured the imagination of the media organisations was the story behind the story. As I saw it, I had done nothing more than I usually did – working out a way of safely getting as close to the action as possible, and then doing it. In the media, though, I was being represented as some kind of nutter-adventurer-hero.

Harold, too, was quite unprepared for his new status as international daredevil celebrity. Everyone seemed to think the flight was a big deal, even though Australians such as Harold fly Cessnas over the outback – as hostile an environment as the open ocean – every day of the week without any fuss.

But Harold and I had been the first to Tikopia, five days after the storm, and questions began to be asked about how a news photographer could get to a disaster area before any kind of relief effort – and what exactly was being done for the Tikopians by the governments of the region?

By the time I got back to my hotel, the phones were running hot. They were running so hot, in fact, that the entire telephone system on Santo Island crashed. So, too, did my website, which received over a million hits in a short space of time. One of the organisations which managed to get hold of me was the Australian office of the giant French magazine *Paris-Match*. They wanted me to take a helicopter to Tikopia, and they didn't care how much it cost. That suited me just fine. Our glimpses from the air had given me a strong desire to see what things were like on the ground. Whereas I had thought I'd be sending a story to the world about death and mayhem, I now sensed there was an even better story to be told – how a good number of Tikopians had managed to survive one of the most destructive forces of nature.

I contacted Phil Cotter, who owns the picturesquely named Helikopta Kompani of Vanuatu, and arranged to charter him and his machine to fly me out to the island. He had heard, of course, about our fixed-wing escapade, and was keen to have a look at the island himself. This time I wanted to fill any spare room in the machine with aid supplies. I'm no anthropologist, but I knew enough about Polynesian culture to know you didn't just show up on someone's patch and take pictures without giving them something in return. For the Tikopians, as for most Pacific cultures, every human interaction was a form of exchange, and had to be as fair as possible. And although an Australian airforce Orion had since overflown Tikopia and reported that people seemed to be happily fishing and going about their daily lives, I was pretty sure from what I had seen that the need for supplies would be desperate.

We arranged for a radio, first-aid equipment, water and food. I asked around and gave a lot of thought to what kind of food would be suitable. Ideally, it would be light to carry, nutritious and easily cooked. As a fan of two-minute noodles on various tramping and adventuring expeditions, I felt they were the logical choice, and I was assured the Tikopians would be happy with them.

So on January 4 Phil Cotter and I, together with a local interpreter, a couple of drums of aviation fuel and a helicopter-load of two-minute noodles, lifted off from Espiritu Santo. Phil dropped the nose of the machine and we swung north towards the island of Vanua Lava. Our plan was essentially the same as it had been in the fixed-wing plane, namely to get as far north as possible and refuel from the drums that we were, against regulations and the rules of prudent practice, carrying. We had one advantage in the chopper: we would be able to refuel on the ground in Tikopia.

The flight was a little longer than in the faster Cessna, but there was more to see. As usually happens after a cyclone has been through, the weather had cleared up beautifully. We set down on the beach at Tikopia in glorious sunshine. From the ruins of the village and the bush surged a huge crowd of men, women and children. They swamped us, laughing, smiling, highly excited at receiving their first proper contact with the

Tikopians on their remote island in the Solomons are well-used to cyclones, and their houses have traditionally been built extremely low to the ground. Even so, few of the structures survived Zoë's battering. The island's people are Polynesian and have an advanced system of sustainable agriculture – a fact that led them to be cited in Jared Diamond's 2005 book *Collapse* as an example of a successful society.

outside world since the storm seven days earlier. One of the first questions we asked was how many had died. The answer was hard to believe: none.

As we doled out the noodles and fresh water, I filmed and took photographs and we listened to their stories. The Tikopians, despite being under the political auspices of the Melanesian Solomon Islands, are one of the most remote communities of Polynesians. They came to their island around 3000 years ago, and had seen a few storms in that time. They readily recognised the warning signs – streaks of cirrus, light clouds high in the sky, larger-than-usual waves and thickening cloud banks – and took measures to ride them out. The volcanic interior of the island is riddled with caves, which they stocked with food and sheltered in when things cut up rough. Even their architecture, while primitive in terms of materials – coconut-palm timbers and sago thatch – is cyclone-friendly, being low to the ground, and with tiny entranceways to minimise the purchase hurricane-force winds could get on them. You need only look at Florida, Port Hedland, or anywhere else in the affluent West with a sea view and a high incidence of cyclonic activity, to see that all the modern, high-tech building materials in the world are no substitute for simple common sense.

That said, the Tikopians were now in dire straits. Their crops were gone, their fresh water had been contaminated by the huge waves which had ridden over the entire low-lying part of the island, and the communal pit they used for their sanitary arrangements had been swamped, and was now threatening to spread disease. Our two-minute noodles were welcomed with joy, but one helicopter-load was never going to go around 1500 people.

The gratitude with which we were received was incredibly touching, and it was hard to leave without being able to give the Tikopians definite answers about what was being done to assist them. When we got back to Vanuatu, however, I made sure I included a piece in my broadcast criticising the lack of an organised response to the disaster. Like the original story, this was taken up around the world. The Australian and New Zealand governments found themselves answering questions in parliament. The Solomon Islands government was pilloried by its own people, and the pitiful aid effort it was mounting – dispatch of a patrol boat from Honiara was being disrupted by striking seamen who wanted a bonus before they would accept the mission – became an international scandal.

The Storm That Wouldn't Die – Severe Tropical Cyclone Ingrid, Australia

Darwin, Australia's northernmost city, is haunted by the memory of Cyclone Tracy, which hit on Christmas Day 1974, causing widespread damage and killing 65 people. While the warm seas around the top of Australia – the Coral, Arafura and Timor – present ideal spawning grounds for big, juicy tropical cyclones, few ever reach Darwin. From the Coral Sea, they tend to track down to bother north Queensland. The occasional vagrant which does head west is usually depleted by its interaction with the Cape York Peninsula or Arnhem Land. And those which form to the north of the city tend to scribe an arc west and south, hitting the coastline of Western Australia from Cape Londonderry to Exmouth.

The eye of Severe Tropical Cyclone Ingrid lying north-east of Cairns, northern Queensland, March 7, 2005. NOAA

The only thing certain about tropical cyclones, though, is that nothing's certain. Sooner or later, another big storm is bound to hit Darwin. And for a few days in early 2005 it looked as though Cyclone Ingrid would be that storm. Ingrid began life on March 3 as a tropical low, north of the Gulf of Carpentaria. The Australia Bureau of Meteorology kept a close eye on her as she drifted south-east, and so did I. Over the next couple of days, Ingrid meandered about in the Coral Sea, hoovering up all that moist air. By the morning of March 6, she had reached tropical cyclone proportions. By the time she had intensified to Category 5 two days later, she had also done a U-turn and was heading for the mainland. Cairns was first in line, then the isolated communities further north.

It had already been a busy, if frustrating, Pacific cyclone season. I had chased three, Nancy, Meena and Olaf, all of which had rampaged about Samoa and the Cook Islands without causing large-scale damage, or permitting me to get as close to their eyewalls as I would have liked, although I had managed to film the waves of Meena's storm surge wrecking Trader Jack's, a waterfront restaurant in Rarotonga that was rapidly becoming famous for going underwater every time a tropical cyclone happened by.

SEVERE TROPICAL CYCLONE INGRID	
Duration	March 3–17, 2005
Highest sustained wind speed	250 kph / 155 mph
Lowest barometric pressure	924 hPa
Damage	unknown
Casualties	5+
Areas affected	Northern Queensland, Northern Territory, Western Australia

The trouble with storm-chasing in the Pacific is that the islands are small targets. If you plonk yourself on the wrong one, you may as well have stayed at home, because there's no way you can get from one island to the next with all flights and shipping services suspended for the duration. In Australia, on the other hand, you can get about with relative ease until flooding and road closures shut you down. So with a Cat-5 storm set to make landfall in Australia, it looked like my luck was in.

I got in contact with my friend Dr Jonathan Nott of the meteorology department of James Cook University in Cairns, and arranged to meet him on March 9, with a view to driving towards Cape York. I flew into Cairns that day, and although there are airstrips dotted all the way up Cape York, it looked very much as though Cairns would be the last place on the way where we could rent a vehicle. I hired a four-wheel drive, leaning innocently on the counter listening as the staff explained the clauses which voided the insurance if we drove on gravel roads and forded rivers.

It was about 600 kilometres from Cairns to the point where it looked as though Ingrid was going to come ashore. It took us two days to get there.

You'd normally expect to cover such a distance in a single day, but we found ourselves hammering over an increasingly dodgy road, tar seal giving out to gravel almost immediately after Cairns, and then to little more than a slash of red dirt through the bush, itself cut across by river after river. As we got north, we began to experience the effects of Ingrid's outer rain bands, and the river crossings got hairier and hairier. Signs comically advised us to 'Drive to the Prevailing Conditions', while others warned of the danger of kangaroos or bush cattle bursting out of the scrub.

Five hundred kilometres from Cairns, we stayed at a ramshackle road-house in Musgrave, about 50 kilometres inland from Princess Charlotte Bay. Before we got going again in the morning, I checked the latest reports on the internet, and as we headed north I pulled down regular updates by satellite phone. It became clear that we were going to be too late to get to Ingrid's landing-point, which looked like being near Lockhart River, on the east coast of the peninsula. But we would reach the Archer River Roadhouse in time to catch the action there.

We got to the Roadhouse late on March 12. There was nothing much to it: just a low tin building with a bowser outside and a big draughty room inside. A few scruffy men caked in red dust were slouching on stools at the bar, the odd dog was scratching in a corner, and of course lots of beer was going down. There was accommodation available of a rough-and-ready sort. Our hosts, hard-bitten, true-blue Aussies, were extraordinarily hospitable. If anyone present, including them, was anxious about the approach of the storm, or curious about our determination to get in its way, they didn't let us know.

That night was rough but little damage was done. Ingrid, although fierce, was a small storm, and the worst of the winds and rain had passed to the north of us. We learned later that the Aboriginal community at Lockhart River had taken a bit of a beating, but when the cyclone, already weakened to a Category 4, touched land, she seriously ran out of steam crossing the hot, dry expanse of the peninsula.

Strike one for my Cyclone Ingrid chase.

We now had to reverse direction and get out fast, as from what we had seen of the rivers on the way up the intense rainfall in Ingrid's wake would cut us off before too long. Although there were some alarming moments, with dirty brown water washing well up the doors of the vehicle, we made it back to Musgrave safely. After another night there, we got back to Cairns late on March 13. A quick run through a car wash, and a beautifully clean Land Cruiser was returned to a hire company which was none the wiser.

By now Ingrid had drawn a deep breath of the warm, moist and, for a tropical cyclone, invigorating air above the Gulf of Carpentaria. On March 13 she was back to her Cat-5 best and heading for Arnhem Land and a possible encounter with Darwin. Jonathan was out of time, so I caught the flight to Darwin alone. By contrast with the Cape York communities, there was a fair bit of nervousness evident in Darwin. As the locals don't see many tropical cyclones, they're not particularly well-prepared for them in terms of things such as building standards, emergency procedures and shelters. Usually all they have to deal with is heat, which they do by drinking lots of beer. From what I could see, they were applying the same remedy to the oncoming cyclone. I settled back to watch Ingrid bump her way along the Northern Territory coast toward us.

Things were looking good as she grazed Cape Wessel at the north-east tip of Arnhem Land. Darwin went to Yellow Alert, the second highest warning on the scale. After that, though, Ingrid began to track slightly north of due west. I looked for transport options to the remote Cobourg Peninsula and Tiwi Islands. There were airstrips there, I could see that much, but each time it looked as though I had found someone willing to get me into the path of the storm, the local Aboriginal land-owners stymied it. For safety reasons, they wouldn't give me permission to fly over their lands until *after* the storm.

Ingrid hit the Cobourg peninsula at 9 a.m. on March 13, losing energy as she did so and sliding down the Saffir-Simpson scale to three. Darwin got a bit of a lashing from the rain bands and there was an amount of flooding, but little in the way of destructive winds. Instead, the cyclone hit Melville Island in the Tiwi group.

Strike two on my Ingrid chase.

Ingrid skipped over Melville Island preserving much of her organisation, and it took only a few deep breaths of the Timor Sea to restore her to full Cat-5 health on March 15. The storm that just wouldn't lie down and die began to curve quietly into the Joseph Bonaparte Gulf and in towards the coast again. I briefly contemplated flying to the remote settlement of Broome, picking up my specially built cyclone truck (see page 220), and making a dash to see if I could intercept Ingrid as she came ashore in Western Australia. However the distance involved – about 1200 kilometres on poor roads – convinced me it would be better to stay local.

I managed to get on to a flight to Kununurra, a town about 50 kilometres inland, and 40 kilometres from the boundary between the Northern Territory and Western Australia. It was a quick, bumpy ride in a small plane, with the pall of cloud associated with Ingrid visible out the small windows. As soon as I was on the ground, I grabbed the latest on the storm's path from the internet. It looked as though the cyclone was heading for Cape Londonderry and the little Aboriginal community of Kalumburu on the coastal fringe of the Kimberley plateau. I made inquiries. Although no sizeable airfield was shown on the maps, there was apparently a strip at Kalumburu where a light plane could land. I even found a Cessna pilot willing to fly me there, and although he cooled on the idea when he looked at the latest weather radar imagery, I managed to convince him to give it a try.

If the flight into Kununurra had been a bit lumpy, the flight to Kalumburu was like a roller-coaster. Ingrid's outer bands were already having an effect: we could see most of the cells, marked by towering cumulonimbus clouds. These big fluffy clouds are a storm cell's signature, areas of intense convection and all the thunder, lightning and hail that accompanies it. Pilots avoid them where possible as the turbulent up- and down-draughts can hurl an aircraft into the sky, or swat it downwards like a fly. But tropical cyclones consist of hundreds of kilometres of storm cells arranged in concentric circles. We would have a bumpy ride, no matter what we did.

Occasionally we'd blunder from a patch of poor visibility into one of violent turbulence. The aeroplane would shudder, and looking out I would see the wings flexing – just another reminder that much of the danger in my line of work was getting where you're going. For all his nervousness about taking on the storm, the pilot appeared remarkably cool.

When we got there, Kalumburu turned out to be nothing more than a shantytown – a few ramshackle houses and a few more substantial but hardly less dilapidated community buildings. The settlement had started as a Benedictine mission to the Aboriginal people at Drysdale River, and been shifted to its present site in 1927. Hardly anyone ever went there, although today, with the boom in four-wheel-drive ownership, increasing numbers of fishermen are finding their way to Western Australia's 'Diamond Coast' and Kalumburu is the gateway. And as I learned during my stay there, at least one outfit has tried to develop this desolate stretch of coast into a tourist destination.

As we came in over the bush – a stubble of parched shrubs forming a threadbare cover on the red earth, pocked here and there with termite mounds – I discovered new truth in the phrase 'the middle of nowhere'. The long arm of the law had clearly reached this far though: a Holden Jackaroo, blue and red lights flashing, was parked next to the strip. As I got off the plane I was met by a police officer, who informed me that alcohol was totally banned in Kalumburu. I thanked the pilot and instructed him to take off as soon as he had taken care of refuelling. Partly, I didn't want him to get stranded by the arrival of the storm. Mostly, though, I didn't want to give the local authorities an opportunity to tell me to get back on the plane and leave.

The plane duly took off. I was stuck in Kalumburu, for better or for worse.

Kalumburu is about as well-geared to tourism as it is to, say, wine-tasting. There's a caravan park and campground run by the mission, but when I tracked down the proprietor, Father Anscar, he rubbished the idea I might stay there. He predicted that his little cabins wouldn't last five minutes when Ingrid came along, and I have to say that when I saw them I reckoned he was erring on the side of optimism. He directed me instead to the courthouse, a relatively sturdy building, and it was there that I set up my satellite gear to track Ingrid's approach, and prepared to weather her arrival.

Others were also taking a professional interest in Ingrid. Sharing the courthouse with me were two men from Western Australia's State Emergency Service, who had flown up by helicopter to fly relief missions to the outlying communities of this godforsaken part of the country after the storm had gone through. I managed to wheedle a promise out of them to take me along for the ride. In the meantime there was the small matter of surviving and filming the storm. Few cyclones of any note had made landfall in the area, so the locals were pretty sceptical when the SES crew and I warned them how bad it would be. At the same time, they were intensely curious about my satellite gear and the beautiful images of the storm. The biggest problem I had tracking the cyclone was the crowd of people who parked their cars, kids, horses, donkeys and dogs around both me and the satellite dish, blurring or altogether blotting out the signal.

I had already struck it unlucky twice in my attempts to get in Ingrid's way. It now looked as though that was about to change. By late afternoon on March 15, the wind had got up and there was debris flying around. Ingrid was still a Category 5, and packing winds in excess of 280 kilometres per hour. People were taking refuge however and wherever they could, while I went about filming. By nightfall it was too dangerous to be outside. The only illumination was from my headlamp and the camera's floodlight. That little pool was not going to give me enough warning if a bit of roofing iron came flying out of the screaming night. I retired to the courthouse.

At about eleven that night, the storm seemed to change gear. The rain thundered on the roof and the wind began roaring and shrieking demonically. There were heavy crashes and thumps as trees came down on the roof. For the best part of five hours, we listened to the storm and pitied the poor souls in their flimsy tin houses. Studying the track of the cyclone, the SES people were anxious about a place on the coast called, appropriately enough, Faraway Bay. There were two people there, caretakers looking after a newly constructed resort, and that was exactly where Ingrid had come ashore.

The next morning dawned sparklingly clear, as is often the case after a storm has rolled through. In Kalumburu, trees and buildings had been wrecked. The courthouse was festooned with palm fronds, branches and downed trees. The placid little creek near Kalumburu, grandly named King Edward River but usually a thing you could just about spit across, was now a dirty brown torrent nearly a kilometre wide. There was some flooding on the outskirts of the settlement but miraculously, given the power of the storm and the slapdash nature of much of the construction, no one had been seriously hurt or killed.

I joined the SES chopper flight to see what was left of Faraway Bay. We followed the King Edward River down to Napier Broome Bay, and got our first look at the piles of kindling which had been the facilities of the new resort. The caretakers were both there, safe and sound, to greet us when we set down. As a precaution against tropical cyclones, the resort's owners had installed a shipping container set in concrete as a dedicated shelter, and the husband-and-wife team had spent a terrifying but safe night in this box, listening to the resort being torn apart around them.

Ingrid stumbled on, losing energy and organisation, to fizzle out two days later near Wyndham, about 150 kilometres south-east of Kalumburu.

The Faraway Bay resort in the far north of Australia was severely damaged by Cyclone Ingrid. Caretakers Neville Jones and Julie Turnbull sheltered in this shipping container during the cyclone, March 17, 2005.

Her track had been nearly 3000 kilometres long, and she had managed to be the only severe tropical cyclone on record to have hit all three of Australia's cyclone-susceptible states.

Back in Kununurra, I decided to fly to Broome. Being so close – less than two hours by air – I couldn't resist the opportunity to take a look at my cyclone-proofed truck and have a play with it. When I opened the garage, I was greeted by the rank smell of diesel coming from a leak in the fuel tank. I went for a drive anyway, and spent a happy few hours trundling over the red barren landscape around Broome. Back in town, I contacted a mechanic to get the leak fixed. As I drove across town to deliver the truck to him, the engine suddenly died. When the mechanic arrived he declared, after a greasy, sweaty spell of swearing and tinkering, that the fuel filters were clogged with rust flakes, probably from the same part of the tank that had sprung the leak.

This, too, was not good, but there was a silver lining: if I'd chosen to go after Ingrid in the truck, as I'd earlier contemplated, I would have got a maximum of 200 kilometres before the vehicle ground to a halt. And now I know I have a shiny new tank and the cleanest fuel lines in Western Australia, just waiting for the next big storm to come along.

RECORD-BREAKING HURRICANES

Severe tropical storms are known in the Americas as hurricanes, from the Spanish *huracan*. They breed in the warm waters of the Atlantic between the months of May and November, the official hurricane season, but can and do occur outside these months as well. The warm, shallow waters of the Gulf of Mexico create ideal conditions for existing storms to intensify, and as much of North America's oil production facilities – drilling rigs, refineries and storage facilities – are situated in the Gulf, it's relatively easy for a storm to wreak economic havoc. And the Gulf states are populous, which means the potential for loss of life is high too.

The *deadliest* hurricane in history was the nameless storm – nameless because modern systems for tracking and naming storms had not yet been devised – that hit Galveston on the Gulf Coast of Texas on September 8, 1900. The exact number of deaths will never be known, but estimates range between 8000 and 12,000, most through drowning. The storm was only a Category 4.

The *costliest* hurricane in history is Katrina, which led to the flooding of New Orleans in September 2005. Up until then, the most expensive had been Andrew, which devastated Florida and Louisiana in 1992. Katrina eclipsed Andrew many times over. The fourth and sixth most expensive storms on record, Wilma and Rita, also occurred in the 2005 season.

On average, each Atlantic hurricane season sees ten named storms – that is, tropical storms with wind speeds exceeding 63 kilometres per hour

Deadliest Hurricanes in the United States [Mainland]

HURRICANE	LOCATION	YEAR	CATEGORY	FATALITIES
–	Galveston, Texas	1900	4	8000+
–	Lake Okeechobee, Florida	1928	4	1836
Katrina	Louisiana/Mississsippi	2005	4	1200
–	Florida Keys/South Texas	1919	4	600
–	New England	1938	3	600
–	Florida Keys	1935	5	408
Audrey	South-west Louisiana, North Texas	1957	4	390
–	North-east USA	1944	3	390
–	Grand Isle, Louisiana	1909	4	350
–	New Orleans, Louisiana	1915	4	275
–	Galveston, Texas	1915	4	275

Note: In 1953, the US National Weather Service began naming storms. At first, the names were all female, but since 1979 male names have also been used.

Costliest Hurricanes in the United States [Mainland]

HURRICANE	LOCATION	YEAR	CATEGORY	DAMAGE (US$ BILLIONS)
Katrina	Louisiana/Mississippi	2005	5	80.0
Andrew	Florida/Louisiana	1992	5	26.5
Charley	Florida	2004	4	15.0
Wilma	Florida	2005	5	14.4
Ivan	Alabama/Florida	2004	3	14.2
Rita	Texas/Louisiana	2005	5	9.4
Frances	Florida	2004	2	8.9
Hugo	South Carolina	1989	4	7.0
Jeanne	Florida	2004	3	6.9

(39 mph, 34 knots). Six reach hurricane intensity, with wind speeds above 119 kilometres per hour (74 mph, 64 knots). Two reach Category 3 or above. The extraordinary 2005 season was the first time since 1933 that the National Oceanic and Atmospheric Agency (NOAA) ran out of names for storms, and resorted to using letters from the Greek alphabet. There were *28 named storms*. Thirteen reached hurricane intensity, and seven Category 3 or above. Of these, Katrina, Rita and Wilma were Category 5. Rita was the most powerful Atlantic storm on record until eclipsed by Wilma two weeks later.

With climate change, can we expect that more and more powerful storms will hit America? Most scientists say yes. According to conventional wisdom, cyclone seasons go in cycles, and 2005 definitely appeared to represent the 'up' part of the current cycle. Many scientists, though, thought it was not the peak. Belief is growing that global warming, which means higher-than-normal average ocean temperatures, may crank storms further up the scale than they would otherwise have reached. Some scientists also think it will shift the cyclone belt northward, putting cities such as New York and Washington right in the firing line. If this happens, the results could be even more catastrophic.

Given the frequency of hurricanes, you would expect US authorities to be well-prepared, but this has not always been the case. While it had long been known, for example, that a direct hit by a decent-sized storm would destroy New Orleans, the events of September 2005 revealed a tragic lack of foresight. You needed only drive along the Gulf Coast to see the folly: endless strips of large flimsy mansions with picture windows, crowded up to the shore to maximise sea views.

Homeless in America – Hurricane Katrina & Hurricane Rita, USA

Residents walk along what is left of a street in Creole, Louisiana, after the rampage of Hurricane Rita, September 30, 2005. Most of the houses and public buildings in the town of 1500 people were left in ruins by the 290-kilometre-an-hour winds.

From a storm-chaser's point of view, the 2005 Atlantic hurricane season was the best ever. By the time Hurricane Epsilon (named for the sixth letter of the Greek alphabet) formed at the end of the official season, it was the 26th named storm. But the season will forever be remembered for two particular hurricanes which hit the Gulf states in the space of a month. The first, Hurricane Katrina, was not only one of the deadliest storms in history, but far and away the most economically damaging ever to hit the North American mainland. To my chagrin, when she struck I was 8000 kilometres away in

Switzerland, filming the calamitous flooding then affecting much of northern Europe. Helplessly, I watched the events in the Gulf unfold on television.

Katrina had been born when two different storm systems merged over the warm waters of the Atlantic. The remnants of Atlantic Tropical Depression 10 (TD10) combined with TD11 to create a new system, TD12, on August 23. One day later this had intensified sufficiently to be named. By August 25 Katrina had become the fourth hurricane for the year. Two days later, despite a little stumble over Miami, she was a Category 3 storm and the Gulf states realised they were in big trouble.

Katrina just kept on getting bigger, even as she began lining up the city of New Orleans. New Orleans is the largest city in Louisiana, and one which just happens to be built below sea level, protected from all the water that surrounds it – notably Lake Pontchartrain and the Mississippi River – by a system of levees, spillways and mighty pumping systems. It had long been feared that a direct hit from a strong enough storm would see this system fail and the city go underwater. Katrina looked very much like that storm.

The American authorities did what is customary when a big storm threatens: they ordered the entire area to be evacuated. More people were reckoned to be on the move than at any time in American history since the Dust Bowl of the 1930s drove 300,000 people from Oklahoma to the west. Even so, the order proved easier to issue than to execute. Greyhound, the main bus service, and Amtrak, which runs the United States passenger rail service, both ceased services to the area well before the storm made its presence felt. Other public passenger services were swamped; while all United States institutions such as hospitals, homes for the elderly and prisons are required to have an approved evacuation plan in place, it had escaped everyone's attention that, when the need arose, most of the institutions in New Orleans were going to be trying to get their people aboard the same buses. That was never going to work. And private transport was similarly stymied as fuel supplies ran short and congestion blocked exit routes. As a result, a large proportion of the city's population was still in the hurricane's path long after they had been told to get out.

Hurricane Katrina began her mainland rampage a little before one o'clock on the morning of August 29, making landfall on the Mississippi coast near the little town of Buras-Triumph as a Category 4 storm. And it wasn't just the immediate district that was affected: she was a huge storm, nearly 192 kilometres (120 miles) in diameter, so the whole of the Mississippi region, including the coasts of Louisiana and Alabama, were hammered.

By six that morning, the eyewall had reached further into Louisiana, and everyone's worst fears had been realised: the storm surge had breached the levees in several places and New Orleans had been inundated. Many of those who had missed getting out of town were sheltering in the Superdome, a sports stadium. As the destructive winds of the eyewall, measured at 255 kilometres per hour (160 mph) grazed past, they lifted the roof of this structure, causing panic. Other people who had waited in vain for the authorities to put their evacuation plans into action died in their hospital beds or rest-home rooms.

Katrina would have been memorable enough for the economic damage she caused, estimated at between US$60 and $130 billion. But that was only the direct cost. The indirect cost was incalculable. Of the 3000-odd oil rigs in the Gulf of Mexico, 108 were totally wrecked and another 50 severely damaged, causing oil prices in America and abroad to soar.

Katrina would also have been memorable for her human toll: nearly 1300 dead and another 6000 unaccounted for. The fate of many of those people may never be known. It was the third deadliest storm in American history after the one which levelled Galveston, Texas, in 1900, killing between 8000 and 12,000 people, and the 1928 Okeechobee Hurricane, which killed 1836.

But the reason Katrina will stay lodged in most memories is because over the next four days, as the world watched on its television screens, the government of the greatest country on Earth did seemingly nothing to help its citizens as they struggled to save themselves from drowning, starvation and disease, and as law and order completely disintegrated. Because most of the people remaining in New Orleans were African-Americans, the pictures looked like something out of an African disaster zone. The difference was that these were *Americans* holding their children up to the lens and pleading with the world to come to their aid. Members of the international community, used to dipping into their pockets to contribute to relief efforts in the Third World and still reeling from the grisly spectacle of the Asian tsunami, were suddenly confronted with requests to send aid money to the United States of America.

Meanwhile the nearly 60,000 residents of New Orleans who had gathered at the Superdome to await evacuation did just that: wait. With sparse sanitary facilities and dangerously low supplies of food, water and medicine, their plight became desperate. The media got in; the rescue effort did not. Some of the refugees abandoned the Superdome and tried to walk out of the city, but were turned back by National Guardsmen, who had been ordered to protect property by stopping people moving about. Looting was reported to be widespread, although it was more likely a case of desperate people seeking the necessities of life than a free PlayStation or DVD. Accusations of racism grew – black people were 'looters' whereas white people were merely 'looking for food'. So did reports of the angry citizens of New Orleans firing on the police and the National Guard. It became hard to tell what was true or false, and how bad things really were. The only thing you could be certain of was that they were very bad indeed.

HURRICANE KATRINA	
Duration	August 23–31, 2005
Highest sustained wind speed	280 kph / 174 mph
Lowest barometric pressure	902 hPa
Damage	US$80 billion
Casualties	1836+
Areas affected	Bahamas, Cuba, USA (eastern seaboard, especially South Florida, Louisiana, Mississippi, Alabama)

American President George W. Bush was widely criticised for the federal government's apparent paralysis; the point was often made that the government could send thousands of its citizens and military equipment halfway around the world and keep them fed, clothed and watered, yet seemed unable to do anything for stricken people in its own land. To be fair to the president, cooler analysis would show it was the system that was flawed, rather than its commander-in-chief. Divisions and lines of responsibility between various state and federal agencies were unclear: no one knew who was supposed to make decisions, and when.

The people at Pioneer Productions were keen for me to visit New Orleans as soon as I could, and when another mean-looking storm began to spin up in the Atlantic in September I had the perfect excuse. For much of her life, Hurricane Rita promised to be even bigger and meaner than Katrina, and it was feared she was lining up the same devastated part of the Gulf Coast. She was the 17th named storm, the ninth hurricane, the fifth major hurricane, and the second Category 5 hurricane of that calamitous season. She was considered to be the strongest hurricane ever to have entered the Gulf of Mexico.

Satellite image of Hurricane Rita, September 23, 2005, the day before she hit the Texas–Louisiana coast.
NOAA

When it became obvious it was all on with Hurricane Rita, I made a mad dash from New Zealand to the United States, arriving two days before she was due to make landfall. In the target area, chaos had already set in. Interestingly, around the same time Taiwan suffered direct hits from three Category 5 typhoons in eight weeks, but without comparable scenes. The difference was that in places like Taiwan whole cities are not evacuated. Instead, the houses are engineered to suit the rigours of the local conditions, which happen to include severe storms. In the Gulf states of the United States, in contrast, building designs take little account of hurricanes. Even though the houses are right in the middle of the zone, they tend to be built not of concrete, but of wood and other flimsy materials. Consequently, when a big storm brews up, the authorities try to truck everyone out.

When I arrived in Los Angeles, I downloaded an update on Rita's progress from the internet. Texas – and in particular, the area around Galveston – looked as though it was staring down the barrel. Houston airports had closed, so I grabbed a flight to Dallas. When I got to Dallas, however, I learned that planes were going into Houston to evacuate people. As they were going in empty, I wangled my way on to one. Believe me, it's pretty weird being the only person on a 747. The flight crew started doing the safety briefing over the public address system, but after a few deep sighs and some self-conscious giggling, they gave up, sat down and buckled their belts for take-off.

We got into Houston about midnight. On the way in, I could see a great band of white light extending from the city toward Dallas: the headlights of thousands of stationary cars of people trying to get out via the choked interstate highway. The I-45 usually has four lanes in either direction, but for evacuation purposes all eight lanes were now open for northbound traffic only.

In Houston, I met up with TV3 reporter Ingrid Leary who, having been already in the US, had been able to get there ahead of me. She had arranged a rental vehicle, a nice, solid four-wheel-drive, but the biggest, toughest vehicle in the world isn't much use without fuel. There was none to be had: all the gas stations were closed, along with everything else. Supplies of food and bottled water were also doubtful. With around two million people, Houston is the fourth largest city in the US, but it was like something out of a doomsday movie. We had to buy tins of petrol from locals at black-market prices. Who knows where they were getting it from, but one of them offered to sell me an electric fuel pump! We sourced petrol from several people, and prices ranged from US$20 to $40 a gallon. Similarly, we paid about US$200 for a box of groceries that wouldn't have been worth more than $20 in ordinary circumstances.

There were no hotel rooms available: they were all full of refugees from the areas devastated by Hurricane Katrina. We were forced to spend an enlightening night in a Red Cross night shelter for the homeless. All was well until about four in the morning, when a group of wired-looking people barged into our room and spent the remaining hours of the night holding anxious cellphone conversations with people about how low their supplies of anti-psychotic medication were running. The next morning, after little sleep, we learned that Rita had already claimed her first victims: before the hurricane had even made landfall, a bus loaded with elderly passengers had caught fire, killing 24.

After collecting the latest updates from the internet we set out for Beaumont, a town of 113,000 people, about 120 kilometres east of Houston and 20 kilometres inland. I had originally planned to go to Cameron, a town right on the coast, but all indications were that Cameron would go underwater when the storm hit: it was altogether too close to sea level. So Beaumont it was. We settled into Hotel Elegante Beaumont, an overblown, Las Vegas-style place complete with a receptionist who looked like a stripper from the 1970s, with wall-to-wall make-up, big blond hair and a very short skirt. Then we set about gathering 'before' shots.

It was clear we had chosen the right spot: as the day wore on, the winds steadily increased in velocity and the rain began to come down in buckets. Rita was on her way. By the time she made landfall, she had been downgraded to Category 3 but was still highly dangerous. The sheer power of a storm, the wind speeds it's packing, are only part of the equation in determining how dangerous it is. You can get a Cat-5 in some parts of the world, notably South Asia, and be at considerably less risk than in a lesser storm in America. The difference is in the amount of debris flying around. When you have buildings coming apart everywhere, the air is filled with dangerous objects.

As the wind speeds increased with Rita's approach, bits of Beaumont started flying around. Ingrid and I watched – and I filmed – as the entire roof peeled off a modern hotel. Piece by piece, the iron whipped off into the gloom. Our own hotel also suffered a fair bit of damage, including all the windows in the lobby blowing in and glass scattering about, but by then Ingrid and I were out driving around, catching the action. We weren't alone: quite a few American television crews were also on the road. By the time Rita's eye was making landfall between Sabine Pass, Texas, and Johnson's Bayou, Louisiana, at a little after two on the morning of September 24, we had all found our way to a sturdy-looking concrete building, probably a bank, to shelter from the destructive climax. Some of the crews were making forays out into the weather to do pieces to camera, the kind of stunt I'd been doing for years. The difference was that when I did it I wore a hard hat and a life-jacket. It wasn't until a young anchorwoman was collected by a large piece of flying debris – fortunately, she was only slightly injured – that everyone began catching on to the safety requirements.

HURRICANE RITA	
Duration	September 17–26, 2005
Highest sustained wind speed	290 kph / 180 mph
Lowest barometric pressure	895 hPa
Damage	over US$9 billion
Casualties	120
Areas affected	Bahamas, Trinidad, Cuba, Mexico (Yucatan), USA (Louisiana, Texas, Mississippi, Arkansas)

The storm raged most of the night. There were reports of tornadoes being generated in parts of the city. Occasionally we'd see brilliant flashes lighting up sections of the skyline, probably from these tornadoes taking out transformers and substations. Ingrid was not at all happy that we were sharing the

city with marauding twisters, even when I explained that they were random and there was no way of predicting where they'd head! When day broke, the storm began at last to abate, and we moved out into the wrecked city to film. At the destroyed Beaumont airport a helicopter, run by the power company that owned the damaged power lines, was operating. Talking to the pilot, I discovered he was an Australian who had once worked in Fiordland as a deer-shooting pilot and loved New Zealand. This gave me an immediate advantage in getting one of the few spaces available in the helicopter during its allocated flying times. The result was a dramatic photo of strong high-tension power pylons, twisted and contorted by the fury of the wind.

All over the city there was a heavy law-and-order presence, with police and National Guardsmen everywhere, but although there was no shortage of electrical goods and liquor stores with their frontages blown out and their wares open to the street, there was no sign of the kind of pillaging for which Katrina had become infamous. Everywhere we went, people would approach us and offer us food and water and a place to stay if we needed it. Under the circumstances, namely the partial demolition of the place where they lived, the hospitality was heartwarming.

A line of downed power pylons near Port Arthur, Texas, September 25, 2005. An estimated two million people lost electricity as Hurricane Rita tore through the south-east United States. Port Arthur is home to oil refineries and chemical factories. Most of the city's 58,000 residents fled before the hurricane struck, and levees prevented greater damage from storm surge.

But the negative side of American society was also on display. At the Elegante we met a young African-American man, about 25 years old, named Kevin Washington. Kevin had his elderly parents and even more elderly grandparents with him. The family had tried evacuating ahead of Rita's arrival, but had run out of fuel. No one had helped them. Then they had been turned away from several hotels. Kevin believed it was because they were black. If we had any doubts about the truth of this, they were dispelled when we saw the shocked reaction of white people to the assistance we were giving this black family.

The following day, Ingrid and I drove down to Cameron. A couple of miles outside the town we were stopped by a police cordon. I asked the officer who leaned into our window how bad things were. 'Well,' he said, 'if you've seen those pictures on the television of that tsunami, you'll have a pretty good idea what it's like in there.' We found his description sadly accurate: Cameron, a town of 10,000 people, had been virtually destroyed. The combination of the storm surge and high winds had ripped and pounded everything to bits. Hardly a building was still standing. It was clear that if we'd made it there before the storm, we would have lost the car, all our gear and most probably our lives as well.

A car crushed by a large signpost in Beaumont. The city's 110,000 people were officially ordered to evacuate as Hurricane Rita approached; it was two weeks before they could begin to return.

ABOVE
In downtown Creole, Louisiana, the post office lies in ruins. Both Creole and neighbouring town Cameron were virtually wiped out by Rita's 240-kilometre-an-hour winds and storm surges. In Creole, the only public buildings to survive the rampage were a courthouse and an elementary school.

LEFT
The owners of a fabrication company pick through the wreckage of their building in Orange, Texas. Rita destroyed more than two-thirds of the businesses in the town, which escaped even more damage when the storm surge spun at the last minute towards Cameron, Louisiana.

Tornadoes

Do the Twist

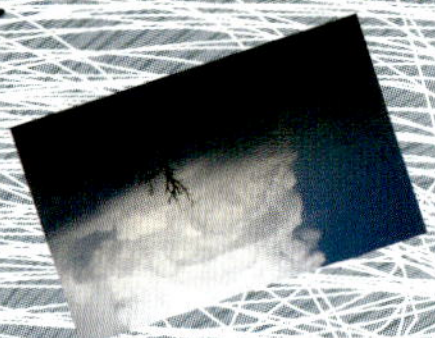

The forces which give rise to tornadoes are broadly similar to those which generate cyclones: heat energy, which causes different masses of air to move, and the Coriolis effect, which causes airstreams to move in a circle.

Tornadoes are very much smaller than cyclones – up to a couple of kilometres across, as against hundreds of kilometres. But they can be no less deadly. A cyclone is like a speeding truck, incredibly dangerous but big, highly visible and relatively predictable. A tornado, by contrast, is more like a landmine. It will appear suddenly, often lasting only a few minutes before vanishing again, and this unpredictability can make it lethal.

Tornadoes – or at least the classical and most dangerous ones – always occur as a consequence of thunderstorm activity. Thunderstorms are generated when cold, dry air masses encounter warmer, moist ones. Areas of intense convection occur, with the warm air rising and carrying its payload of moisture aloft. This process is visible: you see great, fluffy cumulus clouds piling high into the air, often assuming an anvil shape and wreathed about their upper reaches with lateral streaks of cirrus. These are known as thunderheads, or thunder clouds. The warm rising air acquires a positive charge, while the cooler descending air builds up a negative charge. When this electrostatic differential has developed between the upper and lower reaches of a cloud, what you have is a giant battery – an electric cell – hanging in the sky. And this is where the fun begins.

Classic evening supercell, Kansas, May 2000.

Frequently, you'll get storm cells discharging electricity among themselves in the form of cloud-to-cloud lightning bursts. These are seen as flashes or veins of light which do not connect with the ground, and are heard as thunder. Then again, when a hovering storm cell manages to induce a suitably large positive charge in the ground beneath it, it can discharge electricity to the ground, or to whatever prominent object happens to be attached to the ground at that point. That's a lightning strike. It's seen as a brilliant streak of light joining cloud and earth, and is accompanied by thunder.

Another by-product of thunderstorm activity is hail. The moisture in the warm air mass condenses upon contact with the cold air, and most falls as rain. Some, however, is caught in the updraught caused by convection, and carried into the upper reaches of the troposphere, the layer of air which extends from about six to ten kilometres above the Earth. Here it freezes. Much of this frozen material then falls as small ice crystals, which revert to liquid form before they reach the ground. Some, however, is recycled in the convective updraught. Further moisture freezes on the surface of the ice crystal, and it grows correspondingly larger. In particularly intense storms, hailstones can reach the size of marbles or, more rarely, golf balls or, more rarely again, grapefruit. Eventually, they will be heavy enough to defeat the updraught and fall to earth.

And meanwhile, any thunderstorm possesses the potential to generate a tornado. Within a storm cell, two columns of air arise, one ascending, one descending. The Coriolis effect ensures that as the air ascends or descends it rotates. In the northern hemisphere, ascending air rotates clockwise and descending air counterclockwise. In the southern hemisphere, it's the other way around.

Occasionally, the pressure drop at the centre of the rising air mass will produce an intense vortex, which aggravates the rotation of the air around it and draws up air from ground level to feed its hungry centre. An observer lucky enough to be beneath the cloud at this point will be able to see this process going on; the fabric of the cloud itself rotates, a strange and disturbing sight. There's a message in rotating clouds: find yourself a hole in the ground and hide. As the rotation accelerates, the cloud begins to extend downward in a funnel shape. Provided conditions remain favourable to the intensification of the convection, this funnel will drop lower and lower. Eventually it will touch the ground, causing a cloud of whirling debris and detritus, which may change the colour of the funnel from light grey to black. The noise has been compared to low-flying jets, a massed choir of freight trains, or a city-sized swarm of bees.

Some tornadoes are thin threads, just a few metres across. Others are monsters, a couple of kilometres or more in diameter. Most are capable of inflicting major damage. Packing wind speeds of up to 300 kilometres an hour, they can rip buildings to pieces, pick up cars and trucks as though they're toys, and whirl people and animals high into the air and dump them kilometres away. It's not uncommon, after a tornado has passed over a farm, to find it has stripped chickens of their feathers. Someone has even suggested that the intensity of a tornado can be gauged by the proportion of feathers left.

If a tornado crosses a river, lake or tract of sea, becoming a waterspout as it does so, it can draw up whatever is minding its own business in the water – fish, frogs, eels, swimmers – and deposit them kilometres away. Who knows how many religious conversions this phenomenon has triggered in history, but if it starts raining fish or frogs in your town it's most likely the result of the conjunction of a powerful tornado and a distant waterway.

Predicting thunderstorms is relatively easy. Predicting which thunderstorms will generate tornadoes is not. In areas such as the American Midwest, in the collision zone between cold air from the Rocky Mountains and warm air rising from the prairie, it's common for thunderstorm watches or warnings to be accompanied by a tornado watch. People in these areas are usually well-prepared for twisters, with strongly constructed storm cellars beneath their houses, radios which relay tornado alerts whether they're switched on or not, and a keen interest in and understanding of the weather. It's when tornadoes arise in areas not accustomed to them that there's real trouble.

Fujita Tornado Intensity Scale

CATEGORY	DAMAGE	WIND SPEEDS
F0	light	64–115 kph / 40–71 mph
F1	moderate	116–179 kph / 72–111 mph
F2	significant	180–249 kph / 112–155 mph
F3	severe	250–329 kph / 155–204 mph
F4	devastating	330–414 kph / 205–257 mph
F5	incredible	415–508 kph / 258–316 mph

A farmer ploughs his fields, seemingly oblivious to the supercell developing overhead, Colorado, June 2004.

Revenge of the Vortex – Tornado-chasing, USA

It was 1998 and I was ready for a holiday. Hurricane season was drawing to a close. Nothing much was happening around the world in volcano terms. And I'd been thinking for a while about fulfilling another lifelong ambition. There is, in America, a surprisingly large group of people dedicated to chasing extreme weather for kicks. I'd been a reasonably regular viewer of the Storm

Chasers website, and in mid 1997 I'd noticed an exchange of communications sparked by a notice posted on their website by Nancy Bose, a 40-something car saleswoman and mother of three from upstate New York. Nancy reported that she had been intrigued by tornadoes since the age of five when her family, living in Michigan, had been forced to make a dash for their storm shelter because a large tornado was threatening their home. For years after that, the sight of any little black cloud had been enough to start Nancy yelling to her family to take cover. But slowly, over time, her father had taught her about tornadoes – how, where and why they formed. Horror had turned to fascination.

As she got older, Nancy had joined the ranks of the extreme weather nuts. When she saw the movie *Twister*, she wrote, the opening scenes of a family terrorised by a tornado had been like reliving her first childhood encounter, and now she wanted to do what the people in that film did: 'go chase tornadoes'. If there really were crews out there putting themselves in the path of twisters, she would happily serve in any kind of menial capacity on a volunteer basis just to be along for the ride.

Happily for Nancy, the World Wide Web is tailor-made for facilitating contact between people who share obsessions – or, as we obsessives prefer to call them, 'niche interests'. Soon enough, she had a reply. John Bender wrote from northern Illinois, where he had spent many years studying bad weather from as close a range as he could manage. He'd even managed to turn his passion into something useful: he was a volunteer observer for Skywarn, a United States-wide organisation of weather spotters who work with the National Weather Service to inform communities of impending danger from the heavens. Bender had been a member for over 25 years and an instructor of aspiring observers, and was a hotshot at reading meteorological data. If you're serious about catching up with tornadoes, he's the kind of guy you need to have along. While actually getting to the site of a tornado at the same time as the tornado is mostly dumb luck, if you know what you're doing you can maximise your chances.

Bender was happy to arrange a storm-chasing expedition for Nancy Bose, and soon enough they had a third person. Brian McNoldy, a meteorology student at Lycoming College in Pennsylvania, had been a weather addict from an early age, and was majoring in physics and astronomy for a masters degree specialising in atmospherics. And just before I added my name, Dave Ott, a North Dakota farm-boy turned fast-food retailer – he owned two Subway franchises – had come on board. Ott's fascination with tornadoes had begun when, as a teenager, he saw a funnel cloud materialise over his family's house. The tornado had touched earth around 15 kilometres away – right on top of their neighbour's place, which was wiped out. Ott was an active amateur storm-chaser. For him, too, this was a chance to tag along with an expert.

As the 1998 tornado season neared we exchanged hundreds of emails, mostly about the best place to start the chase. This is a bit like looking for a needle in a haystack, except that first you've got to find the haystack. You can work out which parts of North America are most conducive to the formation of the high-energy storm supercells which breed tornadoes. The southern Great Plains region, which stretches from North Dakota to Texas in the weather shadow of the Rocky Mountains, is sometimes referred to as Tornado Alley because of the many large, destructive twisters it experiences each year.

But even then you're talking about a swathe of countryside nearly 5000 kilometres long and 400 kilometres wide. That's a big haystack. And the 'needle' is a fleeting, short-lived thing, lasting a few seconds to several minutes, when it's there at all. There is an average of 100,000 eligible storms in the United States each year; only 1100, or 1 percent, actually breed tornadoes.

We decided to aim for the middle of the belt. We would meet in St Louis, Illinois, rent a Winnebago and then head west, monitoring the internet via mobile phone to hunt for developing storms. I flew in on May 20 and met the others at the airport. They seemed personable enough, a good thing considering I proposed to spend the next ten days in their company, at close quarters and in potentially highly stressful situations. There was certainly no shortage of enthusiasm. John Bender had installed himself as expedition leader, which made sense: he had more experience and know-how than anyone else and owned most of the gear. He had a little car, a Ford Escort, bristling with aerials and crammed with radio and computer gear, and even a home-made lightning strike detector which he was itching to try out.

The group had chosen a name, MESO – Multi-community Environmental Storm Observatory – which gave it a ring of scientific cred. But really everyone was in it for the same reason all storm-chasers – whether amateurs or from university meteorology departments or government agencies – are in it: the thrill of the chase. Nancy Bose had had some T-shirts printed with 'Nice Weather Sucks' and nice weather was what we got at first. It was gloriously fine as we hit Interstate 70, travelling more or less in convoy. None of us was too worried. The heat of the spring sun hitting the huge, reflective surface of the Great Plains always generates hot, moist convection currents. When these interact with the cold air slumping from the Rockies, you get immense, highly unstable thunderheads piling up, the matrices from which funnel clouds can form.

It was my first time in this part of the world. The landscape was featureless, just seas of waist-high wheat as far as the eye could see, relieved only by grain silos and an occasional water tower, windmill or beam pump, nodding as if to assure us there was oil down there. This used to be prairie, a great expanse of grassland roamed by American Indians and huge herds of bison, but today grain has replaced native grasses, and the bison are all but extinct, surviving only in zoos and on the occasional ranch. You get a sense in the drearily beautiful Plains states of just how big America is, especially when you're from a little place like New Zealand, where even on the sprawling Canterbury plains you're only an hour's drive from the nearest mountain.

All our information seemed to suggest that Oklahoma and southern Kansas were the places to be. We headed first to south-central Kansas, and in a medium-sized town called Medicine Lodge we got our first indications both of potential tornado activity and of John Bender's ability to sniff it out. He was monitoring a website which streamed real-time data from a radar site in Goodland, Kansas, close to the Colorado border and several hundred kilometres from where we were. The picture generated by the site indicated the Doppler characteristics of storm activity in its area. Anyone who has done high-school physics will remember the Doppler effect: if you are stationary and a car passes you while sounding its horn, you'll notice the sound changing as the car first approaches and then recedes. The sound rises in pitch as it gets closer, and drops in pitch as it moves further away. Sound travels in waves, and pitch depends on the frequency of these waves.

Deadliest Tornadoes in the United States

DATE	LOCATION	FATALITIES
March 18, 1925	Tri-State (Montana, Illinois, Indiana)	689
May 6, 1840	Natchez, Mississippi	317
May 27, 1896	St Louis, Montana	255
April 5, 1936	Tupelo, Mississippi	216
April 6, 1936	Gainesville, Georgia	203
April 9, 1947	Woodward, Oklahoma	181
April 24, 1908	Amite, Louisiana; Purvis, Mississippi	143
June 12, 1899	New Richmond, Wisconsin	117
June 8, 1953	Flint, Michigan	115
May 11, 1953	Waco, Texas	114
May 18, 1902	Goliad, Texas	114
March 23, 1913	Omaha, Nebraska	103
May 26, 1917	Mattoon, Illinois	101
June 23, 1944	Shinnston, West Virginia	100
April 18, 1880	Marshfield, Montana	99
June 1, 1903	Gainesville, Georgia	98
May 9, 1927	Poplar Bluff, Montana	98
May 10, 1905	Snyder, Oklahoma	97
April 24, 1908	Natchez, Mississippi	91
June 9, 1953	Worcester, Massachusetts	90
April 20, 1920	Starkville, Mississippi; Waco, Alabama	88
June 28, 1924	Lorain, Sandusky, Ohio	85
May 25, 1955	Udall, Kansas	80
September 29, 1927	St Louis, Montana	79
March 27, 1890	Louisville, Kentucky	76

Crudely put, the more sound waves reaching you per second, the higher the pitch. If the source of a sound is steadily approaching you, the effect is the same as the frequency steadily rising, so your perception will be of a sound steadily rising in pitch. The reverse is true of a sound source leaving you. That's the Doppler effect.

Electromagnetic radiation, including radio waves and even visible light, behaves like a wave, and the Doppler effect holds true for this too. Astronomers use 'blue shift' and 'red shift' – the apparent rise or fall of the frequency of light waves – to measure the speed at which celestial bodies are moving towards the Earth or away from it. And meteorologists measure the change in frequency of reflected radio waves – as transmitted, bounced off storm activity and received by a radar – to indicate whether that storm activity is approaching or receding, and how fast. The map Bender was pointing to on his laptop showed receding rain and wind in red, and approaching rain and wind in green. Here and there, there were spots where red and green pixels blinked side by side, indicating patches of weather approaching the radar and other patches receding from it, all in very close proximity. Storm-chasers know this as a 'flying eagle', the signature of a funnel cloud. 'Probably a twister,' Bender said, pointing at one. And sure enough, when we were watching the Weather Channel a little later, a tornado was reported on the ground exactly where he had been indicating. It looked as though we were in the right general area and in the right company.

Next stop was Emporia, 100 kilometres west and south of Kansas City. Bender and McNoldy, using sites on the internet which tracked CAPE (convective available potential energy) data to predict storm genesis, did a good job of getting us to where the action was. By the afternoon of May 24 we were in Ponca City, about 70 kilometres north-west of Tulsa, Oklahoma, watching a dangerous-looking mother-of-pearl sky bank up overhead. It was oppressively hot, 30°C (86°F), with only a gentle, wheat-scented breeze blowing. Dew point – 21°C (70°F) – was ideal for tornado formation. You could feel the prickle of electricity in the air.

The six o'clock bulletin on the Weather Channel showed a band of intense thunderstorm activity heading right for us. On the radar the activity showed up as a set of lurid green splotches, with centres ranging from yellow to cherry-red, depending on the intensity of convection. Tornadoes were expected, the bulletin said. Beginner's luck, they call this: I'd been tornado-hunting for only three days, and it looked like I'd already found some. However, getting a tornado on film was not going to be as easy as walking out the door and pointing the camera. The approaching storm was huge, stretching across a front of 200 kilometres, with mesocyclonic cells forming and dissolving according to completely inscrutable forces and variables.

John Bender, Brian McNoldy and I were looking at the website of the Doppler radar facility at Wichita when we noticed a cell forming a little over 50 kilometres to the north of us. It was small but growing rapidly, indicating strong convection and deep instability. We reckoned it was probably worth a look, so we scrambled our gear together and hit the road, Bose, Bender and I in the Escort and the rest in the Winnebago lumbering behind. John had a laptop on his knee and a cellphone in his hand. An AM radio and a couple of scanners were tuned to stations that might carry advance warning of tornado activity. Every now and then he reverted to the old-fashioned method of gathering weather information, leaning forward and looking out the windscreen toward the west, where the sky was very dirty-looking indeed.

The two-way radio came live and Dave Ott told us he had a fix on the cell due west. His words were given an air of authority by almost continuous bursts of static, indicating electrical discharges associated with a storm. We roared through Newkirk, just south of the Kansas border, and marvelled at how little the population seemed to care about the menace from above.

Over the border, the oncoming traffic had its lights on. The sky ahead was black, grey-green, even purple. We could still see no sign of lightning, but there was plenty of hissing and bursting on the radio. Around 7.30 we reached a small town called Geuda Springs, and heard over the radio that tornadoes were touching down just to the north and west of Wichita, about 70 kilometres away. At this stage we're all thinking, 'Gotta get north as quickly as possible.' But suddenly, about the moment Bender, staring at his laptop, says, 'Wait wait wait,' we all get the feeling we're not going to have to go too much further.

We pull over and the Winnebago pulls in behind us. We're at an intersection on a low hill overlooking wheat fields. The clouds are boiling and seething. They look as if they're about to rip apart with their own weight. And north of us there's a curtain of heavy rain. We are just about right at the heart of a supercharged storm cell. We're not alone: a woman is sitting in a car parked right under the power lines. She tells us not to go too close to the lines because of the danger of lightning strikes. However she makes no move to get away from them herself. There's a theory, which I've never been game to try, that power lines actually shield you from lightning strike, by acting as a conductor. Nevertheless, my instincts say to get as far away as possible from the tallest thing in the landscape when lightning gets going.

And sure enough the cloud bank is brilliantly lit every few seconds by lightning, and the thunder rolls at its heels. We're starting to see cloud-to-cloud and cloud-to-ground lightning. A bolt hits a tree only a few hundred metres away and the tree burns briefly but brightly in the gathering gloom. The thunder is deafening. A bitterly cold breeze is blowing in little fits and starts from the north, carrying the eerie sound of tornado warning sirens from distant towns.

'Probably a twister or two over there,' Bender says, pointing to the sheet of rain. 'They often hide behind the rain.' He points out a band where dark cloud meets light, indicating the presence of the kind of sheer updraught which can generate rotation.

A pickup truck emerges from the gloom, lights blazing, and pulls up alongside us. The wild-eyed bloke inside – a local storm-chaser – reckons funnel clouds have been seen just beyond the rain, only ten kilometres north, and part of a storm cell beating a path southward. His advice? Get the hell out of here. There's a tornado shelter in Geuda Springs, a couple of kilometres away. As he talks, the first big, heavy drops of rain hit the blacktop.

Suddenly, the warning siren at Geuda Springs starts howling. It really is time to go.

When we get to the town, the residents are mostly crammed into the shelter, but there's at least one who isn't. The local police chief is right on hand when Nancy Bose, having somehow managed to get lost in a town with only two streets, is speeding back the way she's come, to try and link up with the Winnebago again. With thunder and lightning, and twisters threatening, the pride of Kansas law enforcement proceeds to write her a ticket.

We make contact again just outside town. By now the rain is hammering down, driven by savage gusts of wind which rock our big mobile home on its suspension like a toy. The radio is going berserk with reports of tornadoes on the ground to our left and right, and ironically, just west of Ponca City where we've come from, a twister, F3 on the Fujita Scale and hundreds of metres wide, is smashing up half a dozen homes and assorted farm buildings.

By the time the conditions force us to pull over and wait out the storm, there are funnel clouds everywhere. We just can't see them. We huddle in the bouncing, bucking Winnebago.

By nine o'clock it's safe to go, so we head back across the border to Ponca City. Even here, we're not out of the woods. The storm is blasting the motel, and the lightning puts on a show which we watch until two in the morning.

The next morning we talked it all over, still buzzing from the thrill. Bender and McNoldy revealed we'd broken all the rules of safe tornado chasing – yes, there are such things – by heading too directly for the centre of the storm cell. There was a strong chance a funnel cloud could have formed right above us, and the first we would have known about it would have been when the world around us began taking off and whirling around. And that might have been the last thing we ever saw.

McNoldy reckoned he had spotted a funnel cloud when a lightning strike backlit the veil of rain to our north. Nancy Bose had been shooting with an ancient digital video camera. When we looked at the footage a little later on, sure enough, if you froze the frame at the moment of the lightning burst there was a funnel all right, stretching from the roiling cloud to the ground. Bose was shiny-eyed with excitement, swearing she'd remember the experience until she was 90. The rest of us were pretty pleased with ourselves as well. We had got to the right place at the right time, and best of all we had got out too, completely intact apart from a speeding ticket. It hardly diminished our excitement to learn that the five twisters which passed over and around us were tiddlers – F0 and packing only gale-force winds. Certainly there was no sign of damage when, after dropping into the Geuda Springs Police Station, we drove back to inspect the area.

Having established there was nothing to see, we set out on the trail of the storm again. It took us south to Ada, around 80 kilometres south-east of Oklahoma City, for Monday night, and on to Mount Pleasant, 175 kilometres east of Dallas, Texas on Tuesday. We found plenty of filthy weather, but nothing with a twister aboard. And then, as so often happens, the system we'd been following ran out of steam. It was Nebraska's turn to cop severe weather, well out of our range.

On the off chance that something else would blow up before we had to go our separate ways, we stayed in Tornado Alley, driving all the way back up north again to the city of Norman, where the National Weather Service's Storm Prediction Center is situated. We did the tour, *de rigueur* for storm-chasers. It was interesting, but not half as interesting as being in the middle of the real thing and seeing the aftermath. Next day we drove to Lamont, Oklahoma, which had suffered a near miss from a F3 twister on that action-packed Sunday night. Generating winds of up to 320 kilometres an hour, the tornado could have torn the town to pieces, but fortunately it had stayed a little to the north, amusing itself destroying farms instead. On its swathe of destruction ten kilometres long and nearly a kilometre wide, it had broken up brick buildings, picked up cars and dropped them, and plucked a trailer home neatly from the blocks on which it was mounted – leaving behind the wooden porch – and turned it to matchwood. That gave us some idea of the forces we were flirting with.

From Lamont, we crossed back into Kansas and checked into a motel in Emporia. The last hurrah of my first storm-chasing trip came when John Bender woke us early on Friday to see whether we were interested in trying to catch up with a storm front sweeping toward Kansas City, Missouri. It was on our route back to St Louis so we all reckoned it was worth a try. The chase was on again. We drove for three furious hours before realising the storm was both too quick for us and not likely to put down any funnels anyway.

Back in St Louis that evening, we decided to rent a video. There was no need to take a vote: everyone but me wanted to see *Twister*. Nancy Bose confessed she'd already seen it 12 or 13 times. Brian McNoldy had lost count of the times he'd seen it, but thought it must be around 50. They couldn't quite believe I hadn't seen it at all. That settled it. Well, we watched it. They asked me what I thought. Rubbish, I reckoned. No movie could compare to the real thing: nature puts on the best show.

I flew out of St Louis the following day for New Zealand. On the flight I reflected on the experience. We had driven thousands of kilometres, slept badly and eaten worse for ten days, and got only a sniff of our quarry. Had it all been worth it? Hell, yes.

ABOVE AND OPPOSITE
Force 5 tornado damage, Moore, Oklahoma, May 3, 1999. During this tornado's journey, its damage path was at times 15 kilometres wide. In Moore it narrowed to 400m, intensifying the force. May 1999 was a record-breaking tornado season in the region, with over 70 observed. The statistics were harrowing: in Oklahoma alone, over 40 people were killed, nearly 700 injured, and damage totalled around $1.2 billion.

I joined the same crew in May 1999 for another chase. We found ourselves back at some of our old haunts, same seedy motel rooms, same awful diet of tough steak, bad pizza and mountains of fries. And the experience was much the same, too.

There was plenty of tornadic weather around, and on a couple of occasions we came very close to witnessing tornadoes on the ground. One of these occasions was a stark reminder of the dangers of chasing. We had positioned ourselves uncomfortably close to a rapidly forming supercell. It looked promising on the computer screens – intense convection, rotation. And it looked even better to the naked eye, a black, evil-looking cloud right above us, flickering with lightning and generating half-formed funnel clouds. It was definitely trying to give us a tornado, albeit a little close for our liking. My instinct was to find some sort of cover. I had noticed an irrigation ditch leading to a culvert under the road. That, I figured, would be a good spot to shelter if a baby funnel got big on us all of a sudden – so long as there weren't any snakes in it. I jumped down, and knelt to peer inside. As I put my hand on the ground, there was a loud crackle and my arm was flung into the air. My first thought was a rattlesnake bite. After a moment, though, I smelt ozone and felt a tingle right through my body. I had been electrocuted. The spots in my vision were the afterglow of an arc between my hand and the ground.

I was curious to find out what had happened. Had I been struck by lightning? And, if so, why hadn't I ended up as toast? A meteorologist later told me

one of two things had taken place: either there had been a lightning strike very close by, and there had been enough of a reservoir of charge in the ground to jump back at me when my hand neared the ground; or there had been a build-up of differential charge in the air around me, about to generate a proper lightning strike, which I had defused by taking a lesser hit. Either way, he assured me, I had been close to something big and bad.

I rejoined the MESO crew once more, in 2000, chasing through Kansas, Oklahoma and Nebraska. Their methods and gear had become steadily more sophisticated, and they were a great bunch of people, dedicated, easy to get along with, and as proficient as it's possible to be at storm-chasing. Each time, though, there were more bodies aboard. And as I now knew all too well, tornado-chasing was an inexact science, and there was a big measure of luck in whether you crossed tracks with a twister, or missed it by a matter of minutes or kilometres. You needed to be as flexible and mobile as possible, and it started to irritate me that we would arrive too late to witness a tornado on the ground because one of the crew had been peckish, or needed a comfort stop. And it seemed to take forever to get moving in the mornings. While I managed to increase my portfolio of aftermath shots and interviews – including two with families outside Brady, Nebraska, whose farms had been flattened by an F1 tornado on May 17, 2000 – MESO no longer seemed the best way for me to get the shots of funnel clouds I was now craving.

I finally caught up with a tornado in 2001, my fourth year of trying. I had decided to team up with Bill Tabor, who had been on the MESO chases in 1999 and 2000. On May 16, I flew to Houston and manhandled my 75-kilogram luggage on to a Greyhound bus for the three-hour trip to Austin. Most of the way, a woman named Loretta negotiated on her cellphone with an insurance agent over damage her laptop had sustained when she ran over it in her driveway.

That night Bill and I pored over the latest weather data, trying to work out a route. We set off the next day, with me driving and Bill operating the chasing gear. We had intended to rendezvous with a *tornadosafaris.com* group at Childress, about 600 kilometres away in the Texas Panhandle, but in the event they had decided to divert to a storm cell across the border at Altus, Oklahoma. Meanwhile, our eye was caught by a 'flying eagle' showing up on the radar in the vicinity of Lubbock, at the base of the Panhandle.

We diverted again when it became obvious we weren't going to catch that cell, and there was one closer at hand in Haskell County. Here we got pretty close to the action. There was a promising-looking rain sheet, and what may have been funnel-clouds forming, although we couldn't be sure in the half-dark. Other tornado-chasers claimed to have seen a tornado in the area, but the Storm Prediction Center hadn't confirmed it. After dark, as we drove to the little town of Sagerton where Bill's parents lived, we were treated to a spectacular lightning show by two severe passing thunderstorms. By the time we arrived after two in the morning, we had covered the best part of 1000 kilometres. Bill's parents greeted us with a hot meal – and a warning about treading on rattlesnakes in the backyard.

Two days later – with no severe weather about, and the only excitement having been provided by, you guessed it, a rattlesnake in the yard – we headed for the Panhandle, which looked the most likely place for fun weather. A big stationary front extending from north-west to south-east across the whole region boded well: such a system would enhance the flow into storm systems and greatly increase atmospheric instability. Storm cells were building along the boundary and a tornado watch had been issued. We figured if we could position ourselves somewhere between Childress, Texas and Altus, Oklahoma, we'd be right in the path.

As we headed that way, we heard a report of tornadoes on the ground near Amarillo, a couple of hundred kilometres away. We were on course to intercept the cell which had produced them. As it was, we were encountering plenty of severe weather. But the storms were mostly 'output dominant' – in other words, not conducive to the formation of rotation about an area of intense convection. There were other promising signs, too: the Interstate was getting clogged with all manner of vehicles bristling with whip antennae and satellite dishes.

As we neared Childress we could see the gust front of a severe storm approaching from the north-west – a wall of dark cloud with attendant dust-storm below. We wanted to be on the south-east flank of this system, but it overran us before we could make it. Everything went dark, hail roared on the roof, and I struggled to control the car in hurricane-force winds and film at the same time. On the radar Bill was watching a series of high-speed rain bands, which were crossing the road ahead of us; he was concerned they might be veiling a tornado. We stopped and let the storm pass, then tried to chase it east. It rapidly got away from us, but there were plenty more where it came from. Darkness fell just as we were on the tail of a cell which seemed certain to produce tornadoes. Sure enough, shortly after we'd broken off the chase, tornado warnings were issued. We headed for Wichita Falls, Oklahoma, expecting a good day's chasing the next day.

Overnight, the weather system responsible for the instability was extruded east-west over Texas and Oklahoma, providing a good focus for intense storm activity. The National Weather Service reckoned eastern Oklahoma and the far north of Texas were the places to be. It suggested that in a small area near Ardmore, about 110 kilometres south of Oklahoma City and 130 kilometres west of Wichita Falls, the prevailing combinations of weather ingredients could produce tornadoes up to F5, the most destructive on Earth.

About one in the afternoon, as we were getting close, we could see two cells arising in the target area. Bill thought they would become tornadic very fast. The radar imagery – our mobile hook-ups were functioning perfectly – showed he was right, but the storms were moving east quickly. To chase them, we'd have to take small roads haunted by farmers tootling along in ancient, overloaded pickups, and their wives serenely doing 40 miles per hour in enormous Chevrolet sedans. And we'd be passing through small towns with low speed limits and bored policemen sitting on their radars. Bill thought it unlikely we would catch up with these cells anyway. And the weather conditions amounted to a tornado factory: there would be plenty more along.

We had just given up on the first two cells when the radar indicated another set forming all around us. We hardly needed the gadgetry to tell us this: great, boiling, towering cloud masses were now filling the sky. The speed at which they built up was breathtaking. Here and there, you could see rotation beginning in the dense cloud, and these areas occasionally lowered a little – the precursor to tornado genesis.

The radio was now bringing reports of tornadoes on the ground, and our progress was being seriously impaired by other traffic. Cars were parked at haphazard angles all over the roadside – and in a couple of memorable cases right in the middle of the road itself – and everywhere people were gawking and pointing. You could tell the novice storm-chasers: they were the ones standing on the roofs of their cars, presenting an ideal target for lightning strike. And it wasn't just stationary cars that were in our way. The moving traffic was also chaos, with drivers gazing out at the storm cells, veering all over the road, and jamming on their brakes with no warning.

As I drove, Bill was busy downloading the latest real-time radar information from InterRAD, a private 'weather pro' company. Suddenly the car was filled with the hair-raising howl of a World War II air-raid siren, the InterRAD warning signal for a tornado. This got our attention. We stopped, jumped out and began filming. Several times, a funnel formed and began oozing toward the ground. Several times, we jumped back in the car and raced after the system as it drifted east. Just when it looked as though the funnel would finally reach the ground, we passed a sign saying 'Welcome to McAlester' and another informing us the speed limit was 25 miles per hour (40 kilometres per hour). Everyone else on the road was obediently doing 25, and no amount of raging, shouting or swearing could persuade them to speed up or get out of the way. The progress of the storm cell was now blocked from view by buildings.

By the time we had got clear of the town, the funnel was in the distance, and the road was taking us in the wrong direction. Before we could find a way to get back on its tail, we were engulfed by a heavy rain squall. We could only watch on the radar as our infant tornado drifted further out of reach.

Another cell had formed a county away, and the National Weather Service had already issued a warning that a tornado was on the ground. It was getting dark. This was our last chance for the day, so we turned around and raced towards the new cell. Progress was slow because we were now in the winding roads that twist through the Ouachita Mountains. Soon, though, we could see, coming towards us, a gust front which looked to all intents and purposes like the end of the world. The towering thunderheads, pitch-black beneath, were gilded and bloodied by the setting sun above, a terrible and beautiful sight. We pressed on until Bill said, 'Uh ... I wonder if this is such a hot idea? This looks as though it could turn into the mother of all tornadoes, and we're going to meet it in the dark.' I stopped. We looked at each other. I turned the vehicle around. The hunters had officially become the hunted.

No sooner had we made the decision to try and outrun the storm than it overran us, swift, sure and deadly. Everything was turmoil. The rain battered the windshield – the wipers couldn't cope, and we couldn't see anything – and high winds began cuffing the car about. I wanted to stop and film, but Bill was frantically trying to match the information from his laptop with his GPS to see whether we were in fact being engulfed by a tornado. I suggested he drive while I filmed, but it was not a good time to try and change places. The wind made driving near impossible, whipping the car from one direction to the next. Bill thought he saw a dark shape cross the road ahead, a few hundred metres away. A tornado perhaps? Suddenly, there were two enormous bangs and flashes in the streaming darkness. Two power transformers had exploded. A few seconds later another one, further away and off the roadside, went up too. Something hit the bodywork of the car with a clunk.

Then, as abruptly as it had come, the storm passed on, leaving us to drive shakily to Wilburton, about 100 kilometres from the Arkansas border. Less than an hour after we were at the heart of a world coming apart, we were sitting in a Pizza Hut eating deep-dish pizza.

Bill and I kept chasing over the next week, going as far south as Louisiana, but although we put ourselves in the way of plenty of severe weather, we never again got as close to a tornado. On May 26 I went to Colorado Springs for a breather – if getting caught out in a blizzard and a severe thunderstorm high on Pike's Peak can be called a breather – and on June 1 I rejoined Bill for another crack at getting that elusive tornado shot. We kept at it for two weeks without significant success.

On June 13, exhausted, we set off for Lincoln, Nebraska, more on a wing and a prayer than a precisely calculated route. We hadn't taken the time to gather and analyse the latest data, so we just headed for the area predicted to have the roughest ride from the latest bunch of severe weather. Our luck was in. As we got south, it was evident we were in a zone of very active storm cells. Near Seward, about 40 kilometres from Lincoln, we pulled over to download the latest data from the internet. We were looking at the radar image of a cell which had formed just to the south of us when the radio came alive, announcing a large tornado on the ground beneath that very cell, and at the same moment we both saw on the screen the Doppler signature for a tornado.

Driving towards the cloud, which had the black, saggy belly we had seen dozens of times over the last weary month of chasing, we reached the crest of a small hill to be greeted by an incredible sight. A huge tornado, 400 or 500 metres across, classically formed and quite beautiful, was less than ten kilometres away. We raced toward it. When it became apparent it was going to cross the road just over a kilometre from where we were, we stopped and set up to film. We filmed for several minutes as it crossed the road. There was

Large, classically formed tornado in Nebraska, June 13, 2001. The most dangerous tornadoes are produced by supercells. As these rotating thunderclouds lower they become funnels, finally hitting the ground with disastrous effects. Tornadoes, which rank among the most destructive storms on Earth, can last from several seconds to over an hour, but most dissipate after less than ten minutes.

a fair bit of traffic on the road and some pulled over, but a surprising number of cars trundled merrily on. At the last moment we'd see the brake lights flare and the car would swerve and pull off the road, or do a rapid U-turn. Exactly how drivers failed to see this huge tornado right in front of them, I had no idea.

We filmed until the tornado was suddenly veiled by rain. Darkness followed, and even heavier rain. We packed up and headed away.

The next day, after a rough night as line after line of severe storms passed over, we returned to film our tornado's impact. Several houses had been completely destroyed, the occupants cowering in their storm cellars as their homes were shredded above them. One person told me he had looked up from the basement of his house right into the heart of the tornado as it passed by. Cars had been picked up and dropped in fields 300 metres away, twisted beyond recognition. Combine harvesters, tractors and heavy agricultural machinery had been hurled about and wrecked. The smell of diesel, and ammonia and propane from a nearby tank farm, hung heavily in the air. Trees had been sliced through.

I spent several hours interviewing the people affected. It always amazes me how people in disaster zones can hold on to their sense of dignity and propriety. These people had just lost everything – home, car, income, livestock, pets – and the media were crowding around filming and asking them how they felt. In the middle of it all, one man laid clean mats over the rubble for us to walk on as we moved around filming his partially destroyed house.

A view across the basement of a two-storey house that took a direct hit from a tornado in Seward, Nebraska, June 13, 2001. The house was levelled. The family were in the basement when the tornado hit and survived with minor injuries. Witnesses we interviewed described the sound as like lying on a train track while a freight train ran over the top of you.

So I had finally seen – and filmed – a tornado. Plenty of people would leave it at that, mission accomplished, go looking for something else, but those people wouldn't be real storm-chasers. Tornadoes are like any of the shows nature puts on when she's in the mood: no matter how much of the action and drama you see, you never stop being thrilled by it. And once you've had a taste, you always want more.

Local police race to stop traffic before a tornado crosses the road south-west of Graettinger, Iowa, June 11, 2004. This was one of about 20 tornadoes reported in Iowa that day.

I went chasing with Bill twice more. We got close to tornadoes, but never quite as close as in 2001. I did, however, acquire one of my fondest memories. We were hot on the trail of a storm cell in Oklahoma which was carrying an F3 warning, strong enough to suck the tarmac off the road. We were so hot on the trail, in fact, it looked as though we and the tornado were about to be fighting for the same stretch of highway. Bill, ever the voice of caution, decided we should take cover as best we could. As we parked under an overbridge to wait for the twister to cross the road, two huge semi-trailers passed us, heading right for the spot where all hell was about to break loose, and completely oblivious to all warnings.

We watched the formation of the tornado on the laptop, watched it scribe its arc right across the road, and when we thought it was safe we made a dash to see if we could film it departing. Bill had been right about the wisdom of stopping. Just how right became apparent when we came across the remains of the two trucks scattered all over the carriageway. As we set up our cameras, the drivers were kicking their way out of the mangled remains of their cabs. Neither was injured, although they were both a little banged about and very shaken. They had one question: 'How the hell do you media people get here so fast?'

When the police arrived a few minutes later, they asked us the very same thing.

After the 2003 season, I received an invitation from F5 Tornado Safaris to join its tour groups. The company had a pretty good deal to offer: if I filmed the chase and provided them with an edited highlights package they could sell to their clients, I could travel free. Well, I jumped at this. They're a good crew, with a real nose for violent weather and the right level of organisation needed to drag groups of around eight tornado virgins about the countryside in time to get among the fun. They have a good strike rate. In only my second season with them, we caught up with a set of beautiful tornadoes in Iowa.

Filming an ominous supercell at sunset in Iowa, June 11, 2004.

Tsunami

Wave of Waves

Tsunami is Japanese for 'harbour wave'. Japan is notoriously subject to earthquakes, and thus no stranger to large displacements of water caused by seismic activity, nor the resulting oceanic-style swells that occasionally surge through sheltered waterways. Tsunamis are sometimes called 'tidal waves', but this is misleading as they have nothing to do with lunar tides – although, if the tide is high, the damage caused by an incoming tsunami will be greater.

A tsunami can be generated in a variety of ways. It can come from a sudden catastrophic change in the sea floor, as when it is raised or lowered by an earthquake. It can be the result of an underwater landslide, or even the slump of a large land mass into the sea. Each of these involves water being displaced, and each is due, directly or indirectly, to the movement of tectonic plates. However, not every earthquake, and not even every *undersea* earthquake, produces a tsunami: a very large displacement of water is required.

A tsunami takes the form of a number of waves. These behave much like any other waves, moving away in a ring from the epicentre of the event. The difference between a tsunami and a big wave lies in their energy levels. There's a fair amount of energy in regular waves – ask anyone who's ever wiped out while surfing Pipeline or Waikiki in Hawai'i – but there's an unbelievable amount in tsunamis. A tsunami is, after all, the energy of an earthquake – a slab of the Earth's crust literally changing position – hydraulically transmitted by the ocean to the shore. The tsunami which resulted from the famous 1883 eruption of Krakatau in Indonesia is estimated to have been equivalent to a 5000-megatonne hydrogen bomb, or 500 million tonnes of TNT. By comparison, the atomic bomb dropped on Hiroshima in 1945 released the equivalent of only 15,000 tonnes of TNT. And the tsunami which took place in the Indian Ocean on December 26, 2004 was much bigger than the Krakatau event.

OPPOSITE
A horrifying moment as tourists try to flee the first tsunami that rolled towards Hat Rai Lay beach in southern Thailand on December 26, 2004 – part of a catastrophic event now known as the Asian, or Boxing Day, Tsunami. In coastal regions from Indonesia to Sri Lanka, giant waves triggered by a 9.2 undersea earthquake left nearly 230,000 people dead or missing.
AFP/AFP/GETTY IMAGES

Plenty of incredible stories were told in the wake of the Asian tsunami. One was that fishermen and scuba divers didn't even notice all that energy going by, even though it flattened the coastline only a few hundred metres away. This is not remarkable when you think of how waves behave. Offshore from Pipeline, for example, you get a long, smooth ocean swell which will gently lift your boat up a few feet, then just as gently lower it. But get that swell interacting with the shallow coral reef inshore and everything changes. The friction of the seabed slows down the front of the wave, even as the back of it keeps travelling at its original speed. The wave gets compressed, and rears up. The height of the wave makes it unstable, it topples over and breaks.

A tsunami, broadly speaking, behaves the same way, except that everything is scaled up. In the open ocean a tsunami can be innocuous-looking, frequently only a few centimetres high. But when you consider the depth of water below – up to several kilometres – there is a vast volume of water caught up in it. And it travels fast: tsunamis move at hundreds of kilometres per hour, sometimes over 700 kilometres per hour.

It's when a tsunami hits the shore that you notice it. Characteristically, because the wave has been caused by displacement of water far away, events on shore mirror the original disturbance. The hollow created by, say, an underwater subsidence can be transmitted ahead of the wave front, as in the Asian tsunami. What you get then is the unnerving phenomenon of the sea disappearing, exposing the seabed, and often leaving fish and marine animals high and dry. People who don't know better – and that was most of us before this 2004 event – find the opportunity to explore the dry seabed and catch fish with our bare hands irresistible. But the golden rule is that if the water disappears, so should you. If you're on the beach when the sea drains away, run like hell for high ground. Because when the water returns, it returns with a vengeance. The front of the wave slows down, but the rest of it is still travelling at hundreds of kilometres an hour. All that energy goes into piling up the water. The wave can reach the height of 30 or even 40 metres, the equivalent of a ten-storey building. Then it falls over.

A large quantity of water moving at speed is like an avalanche or landslide: it will smash everything in its path. The effect of any wave, and a tsunami follows the same rules, will depend on what it finds when it interacts with land. Diffraction at headlands, breakwaters, points and reefs can cause the wave to change direction and lose energy. A long, gently shelving sea-bottom will dissipate much of its energy before it makes landfall. A long, deep, narrow waterway will experience a sudden filling-up, rather than the appearance of a tall, steep wave.

However, an open, steeply shelving coastline will see the wave rear up suddenly and steeply, and anything adjacent will receive most of the energy in the form of rushing, plunging water. That's where you don't want to be when it happens. That's right where hundreds of thousands of people found themselves on December 26, 2004.

World's Deadliest Tsunamis

YEAR	MAGNITUDE	FATALITIES	PRINCIPAL AREAS AFFECTED
2004	9.0	225,000	Indonesia, Sri Lanka, India, Thailand, Somalia, Maldives
410 BC	unknown	100,000	Crete–Santorini, Ancient Greece
1755	8.5	60,000	Portugal, Morocco
1782	7	40,000	South China Sea
1883	unknown	36,500	Krakatau, Indonesia
1707	8.4	30,000	Tokaido, Nankaido, Japan
1896	7.6	26,360	Sanriku, Japan
1868	8.5	25,674	Northern Chile
1792	6.4	15,030	Kyushu Island, Japan
1771	7.4	13,486	Ryukyu Trench, Japan
1976	8.1	8000	Moro Gulf, Philippines
1941	7.6	5000	Andaman Sea, East Coast, India
1933	8.4	3000	Sanriku, Japan
1998	7.1	2182	Papua New Guinea
1923	7.9	2144	Tokaido, Japan
1946	8.1	1997	Nankaido, Japan
1960	9.5	1260	Central Chile
1992	7.5	1000	Flores Sea, Indonesia

The Wave that Shook the World – Asian (Boxing Day) Tsunami, Thailand

The remains of a hotel in Khao Lak in southern Thailand, hit by the full force of the massive tsunami of December 26, 2004. Between 4000 and 5000 people are believed to have died in the beachside town, with the number evenly divided between locals and foreign tourists.

I got home late on the evening of December 26, 2004. I'd been out filming a story about a motor-race meeting which had been held at Auckland's Western Springs speedway in defiance of noise restrictions imposed by the city council. Local residents had been getting uptight about the sound of 20 or 30 guys gunning hotted-up stock cars around a dirt track of an evening, and although the speedway had been there for 75 years, the residents – most of whom had only recently moved into the neighbourhood with the gentrification of the inner city – had won the battle: the speedway was going to close down.

Feelings were running high on both sides, and it was Christmas, slow news season.

Well, that was about to change.

I glanced at the internet to see what was going on in the world beyond the high wire fence around Western Springs. It turned out there had been an earthquake in the Indian Ocean, somewhere off the coast of the Indonesian province of Aceh. It looked as though it was a big one, seven or even eight on the Richter scale, according to preliminary measurements. The initial reports suggested a high death toll. Some of the damage appeared to have been caused by a tsunami. I considered phoning the head of TV3 news to suggest a visit. But in the end I shrugged. It was probably not worth worrying about. I turned in.

By the following morning, though, everything had changed. The magnitude of the quake had been upgraded to nine, and the ensuing tsunami was now believed to have caused loss of life on a massive scale. While I was reading about it on the internet, TV3 phoned and told me to get myself on the ten o'clock plane to Singapore. They wanted me and reporter Bob McNeill in Banda Aceh, the largest town in Indonesia's Aceh province, as soon as possible. I grabbed my camera gear and headed out the door.

By the time we had gone through the formalities at Auckland International Airport, the tsunami was known to have hit a far wider area than first feared. Thailand had been affected, as had Sri Lanka, parts of India and even the coast of Africa. There were stories of whole communities – villages, towns, even cities – having been wiped out. The death toll was now reckoned in the tens of thousands, perhaps far higher. Details were sketchy. The only certainty was that in calamity terms this was major.

ASIAN (BOXING DAY) TSUNAMI	
Date	December 26, 2004
Source	9.3 M_w earthquake 160 kilometres (99 miles) west of Sumatra, Indonesia
Maximum wave height	up to 30 metres / 100 feet
Depth	10 kilometres / 6 miles
Damage	unknown
Casualties	Over 225,000 confirmed

On the way up to Singapore, I mentally ran through a list of everything we'd need in order to film in a disaster area in such a poor part of the world: food and water, of course; independent transport, if it could be arranged; an English-speaking local person who could act as an interpreter; maps… Meanwhile, though, Bob and I speculated that coastal Thailand, if it had been extensively damaged, would appeal to the network as the more newsworthy site. It was a popular destination for Westerners, and if disaster had overtaken holiday spots such as Phi Phi or Phuket you could expect a few New Zealanders to have been caught up in it. It was odds-on we'd be diverted there.

And so it proved. No sooner were we on the ground in the bustling shambles of Singapore Airport than word reached us that we were to get on the earliest flight to Thailand, specifically Phuket. We took our place in the Royal Thai Airlines queue but the signs weren't good. The line was full of dishevelled women and men wearing canvas vests and pushing trollies piled high with aluminium cases – reporters, news anchors, photographers and cameramen. In the time it had taken for a set of three waves to break, Phuket had gone from hot tourist destination to even hotter news story.

Landing at Phuket airport, you'd never have known anything had happened. The airport is several kilometres inland, and apart from a bit of rattling during the quake it had been unaffected. All the talk, though, was of the disaster. No one seemed to be able to find words to describe what we were in for.

We took a taxi to the Best Western hotel. Perched on top of a low hill behind Patong Beach, the hotel had been saved by its altitude and its distance inland. Some of the spectacular amateur camera footage of the tsunami crashing ashore had been filmed by guests looking on horrified from its balconies. Reuters had moved in and installed a satellite uplink, and the massed world media – including us – were using this to get their pictures to the outside world. There would have been close to 500 journalists, reporters, camera operators and photographers in residence. After Bob and I checked in, we opened negotiations with an English-speaking Thai taxi driver, one of a crowd who had beaten a path to the hotel, rightly figuring their services were going to be required.

For our first look at the damage, we followed the rest of the media pack down to the sea. Now we could see for ourselves why it defied description.

Everything was smashed, and overlaid with sand, mud and debris. Emergency workers were picking their way through the mess, and there was already the stench of decay. Bodies don't keep well in the tropics. I'd never been to Thailand for a holiday, but I'd seen pictures of the beautiful coral-sand beaches on television and in glossy magazines and tourist brochures. All that was gone now. The beach inland for hundreds of metres looked like a landfill. It was strewn with palm trees and uprooted vegetation, pieces of buildings, timber, slabs of concrete and an incongruous miscellany of items – televisions, fridges, boats, motorbikes, cars, even trucks. And in the middle of it all were bodies, bodies everywhere, being dug out and swathed in plastic by recovery workers.

Most of the media appeared content to film the area within a couple of kilometres of the hotel. If you were after quick and easy footage to illustrate your story, there was carnage aplenty within a short walk of your air-conditioned bedroom and swimming pool. Best of all from the point of view of the international media, the 'real' news was right there, where the foreigners had died. Bob and I, though, decided to have a look at some of the coastline north of Phuket. On our second day on the ground, we directed our driver out of town. It rapidly became obvious this region had been hit as badly, if not worse. The whole coastal fringe for up to a kilometre inland was flattened. There had been something like 50 hotels along here. Now there was nothing. An eerie silence hung over everything. There were no animals, no birds. At the limits of the fans of mud deposited by the inrushing sea there was a grisly high-water mark of debris. And the further north we got, the worse the view.

Soon enough, we were coming across places where there was nothing but silt between the road and the sea – nothing, that is, apart from the bodies strewn across the mud or partly buried, arms and legs protruding like some kind of nightmarish crop. Few recovery workers had got this far; we were the first outsiders to see, let alone film, these 'body farms', as they came to be called. Here and there, people were picking their way through the fields of grossly distended, discoloured corpses in the hope of finding mothers, fathers, children, husbands, wives. It was futile. Heat-induced decay quickly erases distinguishable human features. Bob was quiet, but then so was I. Bob had reported from war zones and African famines. I had been to places devastated by hurricanes, fatal fires, floods and horrendous motor accidents. I suppose I thought I'd seen it all. I hadn't.

Picking our way through the debris, we found all kinds of possessions: items of clothing, suitcases, backpacks, passports, wallets. We'd pick up a wallet and find it full of credit cards, a driver's licence, cash. There'd be a name and often a photo of the owner: dead, barring some kind of miracle. Later, we heard of media crews who took wallets and passports back to town and gave them to the authorities. Some of the wallets contained substantial sums of money, but there seemed little hope this would ever find its way to the dead person's next of kin.

Perhaps it was the horror of what we were seeing daily in Phuket, but the media – especially the small New Zealand contingent – tended to stick together. Usually, an initially collegial feeling among rival news teams will give way to a competitive urge to get the best, or at least a unique, shot. On this story, rivalry was less evident. We hung about as a group at the hotel, sometimes talking over what we had seen, but more often talking about anything but.

There was no shortage of horror. Because so many foreigners had been involved in the disaster, and so many were unaccounted for, the Thai authorities did their best to set up a system for identifying bodies. Although it would have been prudent to have disposed of the dead as quickly as possible, everything possible was done to store them while the identification process was carried out. Refrigerated containers were brought in, fitted with shelving and stacked with corpses. Plastic-wrapped bodies were laid out on warehouse floors and great piles of ice dumped among them to try to slow the process of decomposition. Bob and I put on surgical caps and masks, plastic ponchos and gumboots and filmed in some of these makeshift morgues. The horribly bloated, plastic-swathed corpses littered the concrete under the glare of fluorescent lighting. White mist swirled from the mounds of ice and hung in a hip-high layer. It was surreal, like something from a science-fiction film. But the hammering of the generators, the slosh of your gumboots through the liquid leaking from the pathetic bundles, and above all the stench, were there to remind you that this was all horribly real.

More bodies were arriving all the time, as they were dug from the debris or surfaced out in the bay and bobbed gently ashore. The chances of anyone recognising them grew slimmer by the hour. After a couple of days, there was no chance at all. Relatives of the missing kept arriving. Two New Zealand families, without real hope in their eyes, prevailed on Bob and me to take a sheaf of battered photos through the morgues to try and identify their loved ones. We had no luck.

Shell-shocked Westerners – survivors, relatives and volunteer victim-identification experts, who had begun arriving in droves – wandered about in surgical masks with hands pressed to their faces. In several places, particularly by the ruins of hotels and resorts, noticeboards had been erected and were now festooned with brightly coloured photographs of missing people, and messages from their relatives. At first the photos were mostly holiday snaps – smiling, happy, healthy-looking people, young, tanned, relaxed. Almost certainly these people were now dead. Later the boards began to feature images from the supply-side: photos of the faces of decomposing corpses, barely human to look at, but posted in case someone might recognise their wife, husband, child, mate or lover.

LEFT
New Zealand television reporter Bob McNeill and I kitted up in surgical gowns to enter an area where more than 2000 bodies lay waiting for identification two days after the 2004 Asian tsunami, Takuapa, southern Thailand.

TOP LEFT
More than a hundred bloated bodies of tsunami victims lie in the open at Takuapa. Forensic teams were carrying out the grim task of trying to identify them.

ABOVE
Vision of hell: a small part of the scene at Takuapa. The photo doesn't even begin to convey what we saw.

Aceh: Another Name for Hell –
Asian (Boxing Day) Tsunami, Indonesia

Dazed Achinese, wearing face masks to combat smell and disease, inspect what's left of downtown Banda Aceh a week after the 2004 tsunami.

During the eight days we were in Thailand the death toll kept growing. We didn't need anyone to tell us that the early estimates had been ridiculously conservative: a good day was when we saw fewer than 5000 corpses.

While there, I was contacted by Pioneer Productions, the company which had produced my *Dangerman* series for Discovery Channel: they wanted me to go to Aceh and shoot a documentary. Richard McNair, by now recovered from his Mount Semeru sickness, flew from Auckland to act as my assistant.

Richard and I set about arranging as much in the way of water and food supplies as we could. Westerners were incredulous when I told them where I was heading. Hadn't I heard that Aceh was in a state of civil war? That foreigners were personae non gratae? The rebels would take one look at us and shoot us, or hack us to bits. I was sceptical about this. By now the media were casting about for new angles on the tsunami, and the possibility of rebels resisting the invasion of foreign aid workers and media was an appealing one. Unfortunately, though, it didn't square with the facts.

My main concern was that I would be taking Richard, and an Australian aid worker who had decided to join us, into danger, to a disaster zone where at the very least there was going to be no kind of infrastructure. I decided it would be best if we had not only our own supplies, but also our own means of getting around. Rather than travel direct to Banda Aceh via one of the relief flights operating, I decided we would fly into Medan on the north-east coast of Sumatra, and once there hire a minivan and driver. That way we could be as self-sufficient as possible – independently mobile and provisioned, and with some means of recharging the camera equipment.

In Medan we crammed every available cavity of the van with food and water, and set out on the 12-hour trip to Banda Aceh. If the first part of the journey was harrowing – the traffic in downtown Medan proving driving was no better in this remote part of the Indonesian archipelago than in Jakarta – the second and longer section was worse. The driver kept nodding off. The van would veer alarmingly towards the shoulder of the road, the outside of corners or oncoming traffic until we yelled at the driver to wake up. I kept demanding furiously that he pull over and let one of us drive, but he pretended not to understand. Somehow we made it, but despite all the hazards created by the natural disaster, and the dire warnings about the political situation, that drive turned out to be the most dangerous part of the trip.

As we got further west, we began to see signs of the tsunami's impact. Aceh is no pleasure spot at the best of times, but in a few short minutes on December 26 it had been transformed it into something close to hell on Earth. When we reached the city of Banda Aceh itself and saw that two-thirds had been completely levelled, we realised that the final toll of the disaster would probably never be known. It was as though an atomic bomb had hit. Building standards in this part of the world were generally not high, but even if they had

All that remains of a residential suburb of Banda Aceh, January 10, 2005. Aceh was the closest city to the epicentre of the massive earthquake that set off one of the world's deadliest-ever tsunamis. 61,000 of the residents were killed.

been few buildings would have withstood the forces that had been unleashed. Here and there, a recognisable structure protruded from the rubble. One imposing mosque seemed practically unscathed, while everything around it had gone.

We managed to find a relatively intact house in which we could base ourselves. It had once been opulent, probably the mansion of the governor or some other dignitary. Now, though, it was derelict. The walls were cracked, and plaster from the ceilings had been brought down by the earthquake which had preceded the tsunami – and by the aftershocks, which still rattled the place from time to time. I figured if the building had withstood the main quake, it would stand up to anything smaller. That was a comfort.

With the bridge gone, we helped locals build a rudimentary barge to cross a river south of Banda Aceh, on the road to the cement works at Lhok Nga, January 10, 2005.

We had been in Banda Aceh barely four hours when things began falling apart. Richard suddenly announced that his girlfriend back in Auckland was 'freaking out' and would leave him if he didn't return to New Zealand straight away. He promptly left and got on the next plane out, leaving me with no assistant. In desperation, I gave the Australian aid worker, Daniel McKeag, a crash course in filming and appointed him to the job.

Once all this had taken place, Daniel and I set out to explore what was left of the countryside. The few media in town were, again, sticking to the easy shots. Most seemed to be stopping on the banks of a river just out of town where the bridge was down, and taking long shots of what lay beyond. That made our next move obvious: get across the river. We found a group of locals trying to put together some kind of raft, and calling on rusty boy-scout skills gave them a hand. Soon enough, with bits of timber, rope and oil drums, we managed to cobble something together: nothing you'd put to sea on, but good enough to float us across the lazily flowing water. Better yet, with the anxious locals looking on, we proved it was capable of taking a motorbike across as well. That was going to save us a bit of walking in the searing tropical heat.

Once on the other side, while our partners in the raft-making enterprise trudged off down the road and in sundry other directions seeking news of their homes and relatives, we loaded up the motorbike and I set off for the strange tableau I had seen in the middle distance – the ruins of a cement works, and a fairly sizeable ship lying high and dry on its side. It turned out to be about five kilometres away, and when I got there I couldn't believe my eyes. There were in fact *three* ships. The other two were parked on a relatively even keel in the middle of the road. Beyond them, the cement works had been smashed up to three storeys above the ground by the onslaught of the water. All around the site there were high cliffs, and a tide mark was clearly visible 30 metres up. The whole place was utterly silent, like a scene from a post-apocalypse movie.

ABOVE AND BELOW

The awesome forces of nature. I could not believe my eyes when I came upon these two ships, which had been hurled ashore by the tsunami and were now straddling the main road at Lhok Nga. A third lay on its side in the sea like a broken bath toy.

I offloaded the gear and gunned the bike back down the road to pick up another load. Daniel asked me what I'd found. I opened my mouth to answer, then shrugged: 'You'll have to see for yourself,' was all I could get out. On the last trip, Daniel came with me. We wandered about filming and I did a piece to camera, standing in the road, dwarfed by the stranded ships. The ships' huge hulls were a powerful visual image of a time when, for a few terrifying minutes, the laws of physics were suspended and the sea changed places with the land.

As we made our way further up the road, we encountered survivors trudging through the silt. Their faces wore the blank, glazed look of people who had witnessed the unimaginable. They were all desperately thirsty and we handed out the bottles of water we had with us, figuring we could last a few hours until we got back to the van. When we returned, however, we found our driver had also been handing out bottles of water. There was none left. It wasn't until we arrived back at our base that we got to slake our own thirst. Water and other essentials were available in the town, albeit in short supply and at hugely inflated prices. Contaminated water is the greatest danger in disaster areas: even a tiny quantity taken through the mouth while you're showering can give you a range of nasty bugs.

On the road to Lhok Nga, I encountered this group of refugees, who had walked for six days without food from a town about 160 kilometres away. I alerted Indonesian aid agencies to their plight, but as these people were supposedly 'rebels', no interest was expressed in helping them. Aceh has harboured an independence movement, Free Aceh, for 30 years. After the tsunami, the movement declared a cessation of hostilities in an attempt to allow aid to get through.

A crew from one of the big Australian networks had arrived, and had reached the same conclusion we had about the house we were using. We were competitors, both dead-set on getting the best, most exclusive footage possible, but each had something the other wanted. They were eyeing up our comparatively comfortable digs; I was eyeing up their diesel-powered generator. Negotiations began, and we struck a deal. They could share the house. We could use their generator.

After we had spent two days filming the devastation around Banda Aceh and the seemingly feeble official response, an aid agency offered us a ride down the coast on its helicopter. The relief effort, which featured every major aid agency in the world and quite a few I'd never heard of, had got underway within hours of the disaster and by now it was in full swing. The people in these agencies know that pictures of disaster areas are vital to the flow of aid. Indeed, the more pitiful the images beamed into the lounge-rooms of Westerners, the more deeply they will dig into their pockets. So whereas you might suppose that aid workers would regard the media as freeloading vultures, the opposite is the case: they actively assist us to get the kind of pictures that will tug on the heartstrings, and therefore purse-strings, of the affluent.

From the air the impact of the tsunami was plain. Everything adjacent to the coast had been smashed, and as we're not talking tourist country here, with well-built, high-rise hotels, the death toll was very high. Most of the dwellings had been of poor construction and on ground level, so they had simply been swept away. We landed in what was left of a little town named Malabouh on the south-west coast of Sumatra. Malabouh had been annihilated. We were met on the ground by residents desperate for supplies. There was nowhere near enough to go around, but everything was distributed as best it could be. Having delivered its load and us, the helicopter lifted off again, with those on board promising to pick us up again on their next supply mission.

Zulfahli Anwar, one of the survivors, spoke quite good English and agreed to be our interpreter. He had been inland, he told us, and out of harm's way on the fateful morning. He had come back to his home town to see what he could discover about the fate of his family, friends and relatives. We were with Zulfahli as he picked his way through the wreckage. It was like a rubbish dump, with practically every recognisable landmark erased, or overlaid with debris.

From the air, the utter devastation wrought by the tsunami was even more apparent. In these towns between Banda Aceh and Malabouh few buildings had been left standing.

After a bit, he decided the tract of garbage we were traversing was his neighbourhood. Sure enough, he managed to identify his motorbike, now a twisted hunk of metal. Then he recognised something else. I held the camera steady on his face as he explained that he was looking at the doorstep to his house. That was it: everything but the doorstep was gone.

He was carrying a list of people, which he showed to other survivors picking through the rubble. There were around 30 names on the list. Zulfahli listened to the stories of his fellow townspeople in their rapid singsong language and then translated for us. When, after these conversations, he drew lines through names, no translation was required.

The helicopter didn't return for another three days, during which time our food and water ran seriously low and it began to look as though we were going to share the plight of the locals. We were as delighted as they were when we finally heard the throbbing of the rotor. The noise underlined one of the most unsettling features of the place: the silence.

It was when we were back in Banda Aceh again that I had the two worst moments of the whole experience. One happened when we were driving back from a day's filming. In a field just outside the town, I saw a digger and a large heap of reddish-brown soil in a field next to the road. We had happened along just in time to see a dump truck upending its tray to deliver a few cubic metres of bodies into a great pit which had been gouged out of the earth. With few or no foreigners having been in Aceh when the tsunami hit, no effort was being made to identify bodies; they were just being recovered and buried as swiftly and efficiently as possible in countless mass graves such as this one. I didn't film the scene: I'd had a gutsful of horror by then and just wanted to put the grisly sight behind me.

Now I wish I had stopped and filmed it. The Boxing Day tsunami is destined to be remembered as one of the most significant moments of the twenty-first century, and the mass graves, showing the sheer scale of the loss of life, will be one of its iconic images – like the stick figures in rags at the liberation of Auschwitz, the pyramids of skulls recovered from the killing fields of Pol Pot's Cambodia, and the twin towers of New York's World Trade Center cascading to the ground. It would be one of the few occasions when I left a disaster with regrets over photos not taken.

And the other moment? We had been filming a concrete and iron girder bridge which was festooned with flotsam. The water was now a good three metres below the deck, but the debris was four or five above it, indicating how high the water had raged. The rest of the crew were carrying our gear back to the van and I was following. As I walked I saw a brightly coloured piece of fabric caught up in a clump of rotting sticks and vegetation clotted about the bridge's handrail, and idly yanked on it. It came away, together with its contents – the blackened, bloated body of a dead baby. I blinked at it for a few moments, then put it down on the deck. When I caught up with the others, I asked our driver whether we should do something. Should we bury the body, or at least alert someone to it?

'No,' he said. 'Just leave it.'

OPPOSITE
A mosque and the outline of roads were all that remained of the town of Lhok Nga, Aceh, once a popular destination for surfers from around the world.

Earthquakes

Deadly Collisions

Of all natural disasters, earthquakes are the deadliest and least predictable. The forces involved are indescribably huge, even though they represent only a minute portion of the energy latent in the Earth. Nothing human beings can contrive, right up to the biggest hydrogen bomb, or even all the nuclear weapons ever built, can begin to match the energy released in a moderate-sized earthquake.

For most of human history, the reason for earthquakes remained a mystery. Ancient peoples believed they were the acts of an angry god. This was about as useful as most scientific hypotheses. Even the latest theory – which is only 40 years old – while seeming to explain the geology of earthquakes, has so far failed to produce good methods of predicting them. Thanks to plate tectonics, we can now say with a reasonable degree of assurance *where* earthquakes will happen. What we don't know is *when*.

The theory of plate tectonics is the synthesis of two other scientific theories: continental drift and sea-floor spreading. Geologists had long noted evidence that suggested that the continents, while today separated by tracts of ocean and thousands of miles, had once been joined in a single landmass. In 1912, a geologist named Alfred Wegener proposed that the continents were actually mobile, and drifted on the fluid inner layers of the Earth with the rotation of the planet .

While evidence kept mounting that the continents were formerly contiguous, the theory that it was the Earth's rotation that drove their

When a deadly earthquake – the strongest in a century – struck northern Pakistan, the once-scenic mountain town of Balakot, straddling the Kunhar River, was hard hit. Higher in the Himalayas, many villages rolled down mountainsides. October 22, 2005

motion was rather less convincing. In the 1960s two scientists, Harry Hess and Robert Dietz, independently decided that the motion of parts of the crust – the solid outer layer of the Earth – came about through the continuous production of new crust along the 60,000 kilometres of underwater ridges which run like seams beneath the oceans. Dietz called this 'sea-floor spreading'. According to the theory, the mantle, the layer of molten rock beneath the crust, moved (as all liquids do) in convection currents: hot fluid rose and displaced colder fluid, which sank. At the sub-oceanic ridges, the rising mantle reached the surface and solidified into new swathes of crust, which spread out over the surface of the mantle, pushing the existing crust outward from the ridge. The undersea crust was forced underneath the margins of the continents in a process called subduction.

In the middle of the 1960s, another geologist, John Tuzo Wilson, coined the term 'plates' to describe the independent tracts into which the Earth's crust was divided. In 1967, American geophysicist W. Jason Morgan announced that the Earth's crust consisted of 12 plates, and a month or so later French geophysicist Xavier le Pichon produced a map showing their boundaries. The theory of 'plate tectonics' had arrived.

According to the theory, the Earth's crust is divided into plates, which move independently of one another in response to the convection of the mantle beneath. The motion of plates relative to one another can be 'divergent', 'convergent' or 'transform'. Divergent motion is where plates move away from one another, as in the Mid-Atlantic Ridge. Convergent plates are moving toward one another, and are in collision, such as at the western margin of South America. Transform motion occurs where plates move alongside one another, as at the San Andreas fault in North America.

Seismic disturbances – earthquakes – are the result of the interaction of plates, and most commonly occur at boundaries where the relative motion is convergent or transform. Where plates converge, the boundary may be constructive, pushing the Earth's crust upward as in the great mountain ranges of the world, the Himalayas and the Andes; or it may be destructive, where one or both plates are pushed downward in the process known as subduction. Earthquakes come about at convergent boundaries, where tension builds up in the collision, with the two plates 'stuck' in place until they are suddenly and violently released. The resulting sudden rearrangement of the crust by the movement of the plates is an earthquake.

The same process is at work where plates are grinding alongside one another. Again, they may 'snag' one another and tension build at the site of the snag, until it is suddenly released and the crust rearranges itself. The movements involved may be comparatively small, but on a human scale the energy they release is massive.

Architecture has evolved to the point where some buildings can withstand the motion of the ground on which they are built – but only up to a point. And, of course, seismic building techniques are not practised in the poorer parts of the world. Some of the deadliest earthquakes in history have struck poor, heavily populated areas. The deadliest on record, a jolt of unknown magnitude in Shansi, China in 1556, is believed to have killed 830,000 people. Earthquakes can also cause tsunamis, which represent a destructive force all of their own.

Since 1979, scientists have measured earthquakes according to the Moment Magnitude scale, which ranks them from 0 to 10, where 10 is the most severe. Like its predecessor the Richter scale, the MM Scale is logarithmic – that is, each level is ten times the previous level, so a quake rating 6 on the intensity scale is ten times more powerful than one registering 5. Anything ranking 6 or above is a major quake and will do serious damage. Once you get into the eights and nines, you're in catastrophe country. The Asian tsunami of 2004 was triggered by a quake of magnitude 9.1. The 2005 Kashmir earthquake was measured at 7.6.

World's Deadliest Earthquakes

DATE	LOCATION	FATALITIES	MAGNITUDE
January 23, 1556	Shansi, China	830,000	~8
July 27, 1976	Tangshan, China	255,000	7.5
August 9, 1138	Aleppo, Syria	230,000	unknown
December 26, 2004	Off west coast of northern Sumatra	225,000	9.1
December 22, 856	Damghan, Iran	200,000	unknown
May 22, 1927	Near Xining, Tsinghai, China	200,000	7.9
December 16, 1920	Gansu, China	200,000	7.8
March 23, 893	Ardabil, Iran	150,000	unknown
September 1, 1923	Kwanto, Japan	143,000	7.9
October 5, 1948	Ashgabat, Turkmenistan, USSR	110,000	7.3
December 28, 1908	Messina, Italy	70,000–100,000	7.2
September 1290	Chihli, China	100,000	unknown
October 8, 2005	Kashmir	90,000	7.6
November 1667	Shemakha, Caucasia	80,000	unknown
November 18, 1727	Tabriz, Iran	77,000	unknown
December 25, 1932	Gansu, China	70,000	7.6
November 1, 1755	Lisbon, Portugal	70,000	8.7
May 31, 1970	Peru	66,000	7.9
May 30, 1935	Quetta, Pakistan	30,000–60,000	7.5
January 11, 1693	Sicily, Italy	60,000	unknown
1268	Silicia, Asia Minor	60,000	unknown
June 20, 1990	Iran	50,000	7.7
February 4, 1783	Calabria, Italy	50,000	unknown

The main street of Balakot a week after the earthquake. Every school in the town collapsed and almost an entire generation of children was wiped out.

Moving Mountains – Kashmir Earthquake

On my way back from filming Hurricane Rita and the aftermath of Hurricane Katrina, I picked up a copy of **The Los Angeles Times** *during an airport stopover. On the front page was a photograph of a collapsed ten-storey building somewhere in Islamabad, Pakistan. There had been an earthquake, a big one.* The worst-affected areas seemed to be in and around Kashmir. Damage was extensive and the death toll was likely to be high.

That looks pretty interesting, I thought. I wouldn't be surprised if I end up heading over there soon. No sooner had I turned on my cellphone after disembarking at Auckland International Airport than Pioneer Productions phoned from London. Had I heard about the earthquake? They wanted to do a documentary. Would I please get up there as soon as humanly possible.

I got in touch with my new assistant, 20-year-old Gareth Hawken, and then wasted no time packing. I had met Gareth through a mutual acquaintance a year earlier and he had expressed an interest in working for me. At that time Richard was still on the job, but later, when Gareth heard about Richard's untimely departure from Banda Aceh, he sent me an email saying, 'Gimme a job.' He had joined the Army Signals Corps but had decided it was not for him. In short order Gareth and I were on a plane bound for Bangkok, where we would complete a few formalities and then connect with a flight for Pakistan.

It's not only the squabbling between India and Pakistan which makes life in Kashmir unsettled. Nor is the Line of Control (the unofficial border along which the two countries' armies face off) the only line on which tensions are constantly building. Kashmir is also haplessly situated on the collision zone of the Indian and Eurasian plates, which 50 million years ago began building the Himalayas. Indeed, the Himalayas are still rising by around five millimetres a year.

The same forces make earthquakes common in the region, but on October 8, 2005, at a little after 8.50 a.m. local time, Kashmir was hit by a particularly big jolt. It was centred about 120 kilometres north-east of Islamabad, at a depth of 26 kilometres. At 7.6 on the Magnitude scale, it was a 'major' earthquake, as big as the Quetta earthquake of 1935 or the San Francisco earthquake of 1906. The timing was unfortunate. It was a school morning: most Kashmiri kids were in their classrooms, while their parents were more than likely having a nap after the pre-dawn meal that Muslims typically have during the religious festival of Ramadan. A good many people, therefore, were indoors. Even buildings designed to withstand earthquakes would have suffered extensive damage in a quake this size. In the less well-constructed buildings which prevail in poor parts of the world such as Kashmir, you wouldn't stand much of a chance.

KASHMIR EARTHQUAKE	
Date	October 8, 2005
Magnitude	7.6 M_w
Damage	unknown
Casualties	90,000+ killed; 106,000 injured
Countries affected	Pakistan, India, Afghanistan

In Bangkok, I set about trying to get a journalist's visa for Pakistan. The authorities were reluctant because my TV3 identity card was six years old. I would need a new one before they would accredit me. There was no point in flying all the way back to New Zealand, so we tracked down a company which specialised in manufacturing convincing identification documentation of all descriptions. The press card was one of the easier jobs they'd tackled: the new one looked even better than the genuine article, and I was given the necessary visa with no further problems – unless you count the heavy fee levied for fast-tracking the process. On arrival in Pakistan I learned that foreign journalists intending to cover the earthquake were being admitted without any special visa. I'd been duped.

I've been to so many countries that arriving somewhere foreign seems as natural as getting off a plane in my hometown. However I had read a lot of scary stories – especially in the American media – about how dicey Pakistan was for Westerners, so I was braced for a tough time. The opposite turned out to be true. Everyone we met was courteous, friendly and welcoming. And whereas you'll go a long way to find an American who knows where New Zealand is, every Pakistani seemed to know exactly where it was, knew that Stephen Fleming was the captain of the cricket team, and had a worthwhile opinion on the latest selections in the side as well. We felt safer in Pakistan than in just about any American city you care to name. There was only one sense in which we felt at risk from Muslims, and that was when we encountered them behind the wheel of motor vehicles. Pakistan probably has road rules, but there was little obvious evidence of them.

On the recommendation of an Australian we met at the airport at Islamabad, we hired a driver who spoke perfect English and made all the arrangements for us. This driver was a maniac, even by Pakistani standards, and after a day or two risking life and limb in the bedlam that is Islamabad's traffic, we dumped him. We had our doubts about the second too. Tall and thin, with a greying beard and piercing eyes, he was the spitting image of Osama bin Laden. Indeed, after a little while, we started calling him Osama, which tickled him mightily. He joked to us that the reason bin Laden had been able to evade capture for so long was that every third man in Pakistan was his double.

It was less than a week since I'd first learned in *The Los Angeles Times* of the earthquake, and seen pictures of the collapsed apartment building, yet here I was leaning out of the window of a building adjacent to the destroyed one and taking my own pictures.

Our destination, though, was Kashmir, which had been hardest hit and virtually cut off from the outside world. There were already journalists in there. We listened to their reports every day, and things sounded bad. Not only was the scale of the destruction distressing, but it seemed Westerners were not welcome in this fraught region, where Hindus and Muslims had long lived in a state of low-level armed conflict over the disputed borders. Nevertheless, we needed to get where the action was. Through our Australian friend we hired a man named Raheem to act as our interpreter. We hit it off with Raheem instantly, and had dinner with his family the night before we left Islamabad.

Osama didn't want to leave Islamabad so we hired another driver, whom we swiftly came to call Pepsi as he drank the stuff by the gallon. By local standards Osama had been a prudent driver. When overtaking, he would allow a margin of error of several feet, rather than the more normal five or six inches. Pepsi was nearer the lunatic variety. On the road north of Islamabad he decided the police car we were following was going too slowly, so he pulled out to overtake it on a blind corner. He roared past the police car and about five others, and when an oncoming car loomed up he simply chopped back into the line of traffic. Gareth and I were certain the police would take a pretty dim view of this manoeuvre, and sure enough we heard the loudspeaker behind us blaring. 'They are saying,' Raheem translated, 'please be using your indicator next time.'

We drove about 100 kilometres north to a small (by the standards of the Indian subcontinent) city called Abbotabad, and stayed at a seedy hotel rather incongruously named the Pearl Continental. Had we known what kind of accommodation we were headed for, we would have lived it up at the Pearl. The next day we passed through a village which, we were told, was home to a large population of refugees from the troubles across the border in Afghanistan. Once again, if we'd been going by the stereotypes in the Western media we would have given this village a wide berth, since everyone knows the Afghans are a race of fundamentalist Islamic extremists. Instead, at the shop where we stopped we were treated like royalty. The proprietor refused to take any money for the goods we wanted to purchase: we were his guests, he told us, and it was his honour to give us whatever we needed. Similarly, whereas we'd been told that Kashmir was a no-go zone for Westerners, we learned from the locals that in spite of its reputation there had been, before the earthquake, a steady tourist trade, and that Westerners had been welcomed.

We drove further north to the town of Balakot, which the international media were describing as being the worst hit. The drive from Abbotabad to Balakot would normally take about four hours but time stretched out: the road was clogged with aid and rescue workers heading into the town, and hordes of survivors leaving. The thing that struck me most was that even ten kilometres out of Balakot there was only scant sign of damage. The reports we had heard about Balakot itself had likened the devastation there to the aftermath of the Asian tsunami, so I was ready to discover they had been exaggerated. But, no, the place was flattened. The only way in which the tsumani had been worse was that there the bodies had been strewn across the landscape. Here, they were buried under tonnes of rubble.

We filmed for the rest of the day and then, with no buildings reliably standing, camped for the night in the forecourt of a service station, the only area we could find that was flat, relatively clear of debris, and not overshadowed by structures which could fall on us in any of the hefty aftershocks still rattling through. The other advantage of our campsite was that someone had rigged up a generator. We were able to use the power supply to the attendant's booth to charge the batteries of our camera and communications equipment.

An elderly man begs for help in the main street of Balakot, October 17, 2005. In April 2006 the Pakistan government decided to rebuild the town on another site after a seismological report declared the area unstable.

Every time there was an aftershock, though, I'd sprint out of the booth. The experts say the last thing you should do in an earthquake is run outside, but I would rather die in the open air than be trapped under fallen masonry. Gareth was impressed by my turn of speed.

We did some more filming the next day as people picked their way pathetically through the rubble using only a few hand tools and their bare hands: there was none of the heavy machinery necessary to stage the kind of fast, large-scale recovery operation that might have saved the lives of people trapped under the collapsed buildings. The most distressing aspect was that many of the buildings being frantically and vainly searched were schools. A whole generation of Kashmiri children has been almost wiped out, and their parents were all around us, weeping and scrabbling at the rubble. Few children could be seen, and most of those we did see had suffered hideous injuries. There was little prospect these children would receive timely medical assistance, and every prospect they would die.

That night we camped in a refugee camp, the only media to do so. Most of the other journalists, including those from the world's leading media companies, were flying in and out of Balakot by helicopter from Islamabad, where they were staying in five-star hotels, erasing memories of the day's horrific scenes with $200 meals and cocktails. The difference wasn't lost on the locals, who, despite their desperate plight, showed us great kindness, constantly asking us if there were anything we needed and offering food and bottled water. Many of the people spoke reasonably good English and gave us lucid descriptions of the moment the quake struck. One woman, who had lost her whole family, described the movement of the earth as having been so violent it was impossible to stand up. As the ground bucked up and down, buildings were shaken to bits. We also heard reports from people who had been crossing a large bridge in the town. The deck of the bridge had flopped up and down like a blanket being tossed, bouncing people about like footballs and sending some over the parapet to their deaths in the river, a long way below.

At a crowded hospital, we were given a tour by exhausted medical staff. Afterwards, they left us to our own devices. Try taking a television camera into a hospital in New Zealand or Australia or the United States and see how far you get, but there were no such problems in Pakistan. When we pushed our way into a critical care ward, not a word was said. While we were there, a quake victim died in front of our eyes. It seemed unreal, like a scene from a television drama. The electrocardiograph began to emit a shrill tone, the trace went flat, and a gang of doctors and nurses rushed to the bedside to try and resuscitate the patient. It was too late. A sheet was pulled over the man's face, his body was taken away and the bed given over to one of the thousands of badly injured people packed into the hospital's foyer and corridors.

As we moved around Balakot and the surrounding areas we kept being pulled over by police officers, who would hold long conversations with Raheem before waving us on. I asked him what it was we were so repeatedly doing wrong. Nothing, he replied. It was just that most of the police vehicle pool had been destroyed and the officers were without transport, so they were flagging down cars to try and get a ride to their stations. Later the Balakot police chief himself, his arm in a sling, introduced himself and told us what had happened when the quake hit. The police station, he said, had collapsed, destroying the vehicles and killing the five staff on duty. He himself had been lucky to get out alive. He was now doing what he could to help the relief effort, but without heavy machinery the situation was hopeless.

After another day or so, most of those on the scene realised the futility of trying to save the buried, and much of the effort turned to trying to save the survivors. Aid was slow to arrive, not only because of the damage to roads and other infrastructure inflicted by the quake, and the shortage of the helicopters, but also because the relief effort was poorly organised and painfully slow to mobilise. People were digging in the rubble just as frantically for usable items as for survivors. We were present when a man managed to extract a torch and a set of batteries from the ruins of a shop. The crowd of people assisting him applauded.

We saw many grisly sights. Excavators scratching through the rubble of buildings would occasionally dig through a buried corpse. On one occasion a head turned up, detached by the machine from a still-trapped torso. As a silent, curious crowd watched, a gang dug about trying to extricate the rest of the remains. The head, meanwhile, was placed on a piece of cloth on the roadside, and the driver's licence found in the corpse's clothing laid next to it. The police kindly cleared the crowd aside so that we could have an unobstructed view of this dismal artefact.

Filming at Muzaffarabad, the capital of Pakistani-administered Kashmir and a flourishing city of 700,000 people until hit by the magnitude 7.6 earthquake on October 8, 2005. GARETH HAWKEN

Near the large town of Muzaffarabad, about 120 kilometres from Islamabad and close to the epicentre of the earthquake, we came across a large concrete suspension bridge that had toppled over. Above it, a huge landslide had occurred, gouging out the hillside. The landslide was still unstable, but to get to a good vantage point from which to film the bridge we had to cross it. The ground was still twitching with aftershocks. On the heftier of these a cloud of dust would rise, and boulders the size of small cars would boom down into the river bed. I was impressed by Gareth's calmness. I had expected that, having been in the army, he would cope with the rugged conditions. But he also seemed to deal well with the hardest bit – the emotional impact. Before you're exposed to this sort of horror, there's no way of knowing how it will affect you and this was about as bad as it gets. But despite what he saw, heard and smelt, he continued to get on with the job. In this line of work, woe is strictly after-hours.

Later we would learn the official statistics. In the initial quake over 50,000 people had been killed immediately. Over the next few days, up to 30,000 more had perished where they lay buried or trapped, and still more had died of complications arising from injuries. The total death toll in Pakistan would reach around 90,000.

As I was writing this, over a month after the earthquake, outside assistance had yet to reach some of the worst-affected places. Many of these are in remote, mountainous regions where access is difficult at the best of times. With passes choked by landslides, bridges down and roads blocked, the best-organised relief effort in the world would struggle to reach those in need.

Muzaffarabad in ruins. Aid was slow to arrive, and weeks after the earthquake less than half the villages in the devastated region had even been visited by the government or aid agencies. Many, like this woman, were in shock and despair. GARETH HAWKEN

And this was not the best-organised relief effort in the world. Partly because of 'disaster fatigue' – the Asian tsunami having depleted the world's financial and emotional resources less than a year earlier – and partly because of the horrendous inefficiency of the authorities controlling the afflicted region, aid barely trickled in. An estimated 3.3 million people were left homeless in Pakistan, most in the mountainous districts of Kashmir.

Winter would arrive two months after the quake, and with snow beginning to afflict the hundreds of thousands who still lacked suitable shelter, there was every chance that the death toll from this disaster would go much, much higher.

Snow on the mountains behind Balakot in Pakistan signalled a new threat to survivors – the onset of winter, October 20, 2005.

As seen on Discovery Channel

Danger

Television has been my ticket to a lifestyle that, even in my wildest boyhood technological and meteorological fantasies, I never dreamed of. Britain's Granada was the first international production house to approach me about using my film footage in documentaries: in 1998 the company included my material on the Mount Ruapehu eruption in its series *Eye of the Storm*, billed as 'Footage of extreme weather and eye-witness accounts of some of the terrors weather can bring.'

Next came Pioneer Productions, with its series *Volcano Detectives*, which screened on Discovery Channel in 2001. Pioneer, a London-based independent company, literally pioneered what is called in the trade 'weather porn' – television documentaries about extreme weather – starting in 1994 with *Electric Skies*, about thunderstorms. The first episode of *Volcano Detectives – Eruption* – used the material John Seach and I gathered on our visits to volcanoes in Italy and Indonesia. The second, *Danger Zone*, featured my footage of the volcanoes of Pacific island group Vanuatu, and the spectacular 2000 eruption of Italy's Mount Etna. *Volcano Detectives* won an award at a film festival in the United States. That was a buzz, but nothing compared to the thrill of getting to all the places and volcanoes I'd always wanted to visit.

Abseiling over the crater of Ethiopia's Erta Ale Volcano on a high wire, February 2004. Beneath the deceptively calm crust is boiling hot lava.

In 2002 Natural History New Zealand was commissioned by National Geographic USA and National Geographic International to produce a television series on storm-chasing. Again, it was a bonus for me: instead of having to risk my own money dashing around the world after storms I would be paid to do so by the company.

I teamed up with New Zealanders Mark Whetu and Rachael Wilson and in September, with the ink hardly dry on our contracts, Cyclone Isidore spun up in Mexico. Isidore made for a good pilot for the series, which was to be called *Storm Shooter*. However, soon after it was made I was approached by Pioneer to create a series called *Dangerman*.

This was the big breakthrough I'd been hoping for. With *Dangerman* I would be able to do whatever I wanted...

By September 2003 I had presented them with a wish list: driving a bulldozer into an erupting volcano; purpose-building a vehicle and filming in the eye of a Cat-5 cyclone; and filming on a container ship ploughing through a hurricane. The last stunt was dropped. We contacted shipping companies, but none were keen to have me aboard a ship in a super-storm. This was not because their ships went around cyclones – they didn't – but because of an aversion to showing the public the risks they took, and the number of containers they lost overboard, simply to take the straightest and cheapest route between ports.

At this point Plan B came into effect: ride the Trans-Siberian Railway and visit the coldest inhabited place on Earth.

THE PHOTOGRAPHY BUSINESS

I first began chasing, photographing and filming extreme natural phenomena for the thrill of it. The first inkling I got that there could be money in it came in the wake of my 1995 escapade on erupting Mount Ruapehu.

There was a nice spin-off, too. Having my footage beamed right around the world drew me to the attention of documentary-makers. Any producer making a documentary about extreme natural phenomena needs pictures, but they're hardly going to go out and shoot the footage themselves. Production houses who specialise in the 'awesome forces of nature' genre usually have extensive archives of footage, and use the same material over and over again. The good news for those of us who shoot the footage in the first place is that we get paid each time it screens. I get a couple of calls a week from production companies asking to use a ten-second bit of footage from Yasur, or a sequence from Ruapehu, or a minute's worth of Cyclone Vance, or any number of other historic shots.

I've been burned only once. In 2003 I received a request from a Californian production house for permission to use one of my film clips of a volcanic eruption in Vanuatu for a documentary. I'd never heard of the company and it wasn't offering much – documentary-makers rarely do – but I said yes. A year or so later, when I happened to see a trailer on television for the reality show *Survivor* and recognised some of my footage, I went back to look at the request. Sure enough, near the bottom of the page in small print was *S9*, standing for *Survivor 9*. I wasn't too happy about that. There's a power of difference between a one-off

Dangerman was not slick. It was only roughly scripted, and the action was often spontaneous and unplanned. The series relied on amazing footage, with my describing how unpleasant and dangerous things were. We wanted to take the viewer on our journey, because getting to these kinds of places is always half the fun. The result was a slightly hokey, travel-cum-adventure series. I liked it. The viewing public seemed to like it too: when it screened in the United Kingdom it rated right up there with *The Bill*, *Big Brother* and *Coronation Street*.

Since then the series has screened all over the world. Every day my email inbox is full, and thousands of people have offered to be my assistant.

documentary and 14 episodes of one of the world's top-rating shows.

Then there is the news media. How have I been ripped off by news organisations? Let me count the ways. News producers will often pull out archive footage to illustrate a story, without attribution or payment to the originator. My worst experience came after I supplied footage of one of my cyclone chases to New Zealand's TV3. TV3 is a member of the CanWest network, and it sent the material by satellite to its Australian sister station, Channel 7. Unfortunately, every other network on the planet had access to that particular feed. My footage was grabbed and featured on the news bulletins of almost every major news organisation in the world, none of whom felt the need to pay me for something they'd extracted from a common feed. I lost around $50,000.

The internet can also be the scene of major rip-offs. Newspapers and magazines will often go into a site like mine, download a picture, photoshop out the identification tag, and run the picture without attribution or permission, let alone payment.

The best defence against this theft of intellectual property lies in the process of gathering the material. There are not many photographers filming these events, and so there is a limited pool of images. All photographers recognise their own images instantly, and can identify the source of most of the rest. I get regular calls from fellow storm-chasers who have seen my film footage or still photographs in use somewhere. Usually this is taking place with my knowledge and consent, but I have lost count of the number of times I have sent an organisation a bill, only to have it confess and meekly pay up.

Eye of the Storm –
Typhoon Dujuan, Taiwan Supertyphoon Maemi, Korea Hurricane Isabel, USA

This satellite image from the Typhoon Warning Center shows Typhoon Dujuan approaching Taiwan at 2330 hours. **X** marks my location in the town of Kenting, September 1, 2003.
NOAA

Our chance to film a decent storm for Dangerman *seemed to have come in late August 2003, when a system spun itself up into a Category 3 cyclone in the north-western Pacific, due east of the Philippines.* Typhoon Dujuan's position and path indicated it would make landfall on a populated coast somewhere. The question was where. Because the Philippines was 15 hours away by air, we were going to have to decide quickly, and run the risk when we got there of finding ourselves stranded far from the action.

I put cameraman Joe Morgan on alert, booked flights to China, Hong Kong, the Philippines and Singapore and sat glued to my monitor. Dujuan – the name means 'azalea' in Mandarin – was named on August 30, and after dithering for a day or so, it set off on a determined course just to the north of west. It seemed that either the northern tip of the Philippines or the southern tip of Taiwan were in for it. We decided to head for Hong Kong. That way, we would be in the immediate vicinity when it became clear which it would be.

In Hong Kong airport I logged on to the internet to pull down the latest data. I studied the threat maps, the analyses of the weather systems likely to influence the storm's path, and the latest predictions based on the various forecasting models. Taiwan seemed to be the most likely point of impact. It would hit hard and soon, perhaps in less than eight hours. We frantically organised flights to Taipei, and once there managed to get on the last connecting flight to Kaoshiung, a major port city on Taiwan's south-west coast.

Here we rented a vehicle. For the purposes of the television show, I went through the motions of selecting a little sedan, walking around it while a rental agency employee checked it for dents, and flashing broad winks to camera. In reality we took a four-wheel-drive, which we stocked with food, water and camera gear, assembled and laid out for quick and easy access. You don't want to be messing around changing lenses or putting together tripods in 200-kilometre-an-hour winds.

As we drove, I stayed in touch with the storm's progress through my laptop, which was connected to my satellite phone. Dujuan was making a beeline for the southern tip of the island. We had our eye on a little town named Kenting, which had a view out on to the Luzon Strait – the 250-kilometre-wide body of water which connects the Philippine and South China Seas. The strait was shortly to become cyclone alley. The only problem was how the hell to get there. The road signs were unintelligible, and even when there was a rare place name in Western script it didn't seem to correspond to anything on the map.

TYPHOON DUJUAN	
Duration	August 27 – September 3, 2005
Highest sustained wind speed	225 kph / 140 mph
Lowest barometric pressure	972.1 hPa
Damage	unknown
Casualties	36
Areas affected	Hong Kong, Taiwan, China (Guandong)

With other storm-chasers feeding me GPS coordinates via the internet, we managed to get to Kenting late in the afternoon. At the boat harbour we filmed fishermen trying to secure their boats – and thus their livelihoods – against the mounting storm surge and strengthening winds. But to my intense disappointment Dujuan looked certain to arrive after dark. We had dinner that evening in a restaurant packed with local people going about their business as though nothing were happening. The response to typhoons in Asia has

The coast of southern Taiwan near Kenting, a popular beach resort town on the edge of Taiwan's first national park, as Typhoon Dujuan approaches. Classified as a Category 4 storm, Dujuan had winds estimated at near 225 kph (140 mph). June to September is the typhoon season in the region.

always struck me as quite different from, say, the American response to hurricanes. If Kenting had been an American town with a Cat-3 storm bearing down on it, everyone would have been evacuated. In Taiwan, people were calmly ordering another round of dumplings, another pot of jasmine tea.

Outside, we saw evidence of the same complacency. Rain was hammering down, and the wind was effortlessly hurling large objects through the air, but a man in a see-through plastic raincoat was steadfastly steering his small 50cc motor scooter as best he could through the howling, streaming night.

As the storm intensified, and the eye of Dujuan drifted by just a few kilometres away to the south, we took refuge in an underground car park, filming the wildly thrashing palm trees from the entrance. The power had failed and the air was full of missiles. The winds were now up around the 200-kilometre-an-hour mark: it was too dangerous to try and get worthwhile footage.

After Typhoon Dujuan's rampage, this highway between the towns of Kenting and Kaoshiung was blocked by dozens of fallen power poles. After leaving two dead, the typhoon went on to batter first Hong Kong and then southern China, where it killed at least 38 people and caused damage to infrastructure estimated at US$240 million.

By morning, we were able to venture out and film the aftermath. The road around the shoreline where we had filmed the night before was gone. Several metres from where the breakwater had been, a high-water mark was littered with flotsam – bits of boats, motorbikes, huge chunks of rock.

This had been Joe Morgan's first storm. Although it had been frustrating, with most of the action happening too late at night to get good film footage, he reckoned he could see the attraction. 'I think I may have been bitten by the bug,' he told the tape. 'I now see why storm-chasers get such a buzz.'

Joe's chance to prove this wasn't idle talk came surprisingly soon afterwards.

On September 12, barely a week after we had returned to Auckland, I was on the phone to him again. There were two promising-looking systems. Supertyphoon Maemi was in exactly the same area as Dujuan had been. Hurricane Isabel, which had been lining up the Caribbean, now looked like menacing Florida. Both were Cat-5 on the Saffir-Simpson scale; they would make Dujuan look like a summer breeze.

By midday, Maemi had set sail north-nor'east for Korea, and looked like our baby. Still at Cat-5, she seemed likely to introduce herself to the people of South Korea as the most powerful storm they had ever experienced. Within hours we were on a flight to Tokyo. On the ground at Narita Airport we met up with Isamu Kitafuji, whom I had met at the Mount Usu Volcanic Observatory two years before, and whom I had alerted to our impending flying visit. Isamu worked at Narita and, while we didn't really need his help, the comic sequence we shot of his attempting to show us where to go to catch a flight to Pusan in the labyrinth that is Narita provided the show with light relief.

SUPERTYPHOON MAEMI	
Duration	September 5–15, 2003
Highest sustained wind speed	272 kph / 169 mph
Lowest barometric pressure	885 hPa
Damage	US$4.1 billion
Casualties	117
Areas affected	Japan (Miyakojima, Okinawa); South Korea; North Korea

We left Tokyo two hours later and were soon in Pusan, on the south-east corner of the Korean peninsula. Pusan, a large harbour city with a population of four million, was just about bang in the path of the approaching Maemi. We hurriedly hired a vehicle and stocked up on food and water. With the storm due to make landfall around midday, we had no time to lose.

We hit the road for Sach'on, a little town on South Korea's southern coast which was just slightly out of the city and would provide us with better filming opportunities. Once again, our greatest challenge was interpreting unintelligible road signs, and coping with crazy traffic.

As we got further south, the sky darkened and the rain and wind intensified. Near Sach'on we ran into great columns of traffic, sprawling across all three lanes and into the safety shoulder. People were getting the hell out of Dodge. South Korea received a few good storms a year, but rarely anything on this scale.

Once again, I relied heavily on my satellite gear. Storm-chasers as far afield as North America (where excitement was building over the looming Isabel) lent a hand with sets of GPS coordinates. By the time we finally got to Sach'on about four hours later the winds had really picked up. We filmed a tall, thin man standing on a street corner, raising his hands defensively and flinching with his whole body as miscellaneous pieces of flying debris whipped past him. As the speakers on my laptop tinkled away with the severe-storm-warning tone, I surveyed the data on the internet. We drove down to the harbour to get shots of the seriously big waves pounding the wharf. Back in town, I recorded a piece to camera, kneeling in the roadway to avoid being blown over.

By late afternoon the winds were nearing 200 kilometres an hour, and the combination of a high tide, Maemi's payload of torrential rain and the storm surge had flooded the town. We were still out there and still rolling. It was like urban warfare: we'd huddle behind the corner of a building until one of us plucked up enough courage to race across the road while the other filmed his progress. While we were sheltering and planning our next move, a taxi pulled up, the driver no doubt convinced we would be desperate for a ride out of town – even at special supertyphoon rates.

At one point Joe braved the wind roaring along the road and ran across, while I filmed. After huddling on the other side, strafed by the rain for a couple of minutes, he hustled back. Seconds later we heard a crack. I swung the camera to where he had been standing, in time to catch a cascade of knife-like shards of glass plunging from a blown-out, fourth-floor window. Very soon after that, the power failed. We got that eerie moment on tape too. This was all great *Dangerman* fodder.

After this, there wasn't much to be done but return to our accommodation and wait for morning. As dawn broke, we drove down to the harbour.

I'd entertained a notion of staying in the vehicle there during the night and filming the storm surge at work. Thankfully I hadn't. The waves had overwhelmed a three-metre breakwater and cast around huge rocks. Wrecked trucks, totalled as though by a speeding freight train, lay about. Throughout the town, houses were uninhabitable – flooded or torn apart by the ferocious winds. We filmed a middle-aged couple sadly picking through the remnants of their belongings, which were floating or submerged in dirty brown water in what had been their living room. Altogether, across the country, 25,000 people were homeless. They were the lucky ones. At least another 117 people had lost their lives.

On the way back to Pusan we saw another incredible sight. The previous day I had spotted a floating hotel in the harbour and wondered whether we could use it for accommodation. In the night the huge ship, still moored at the wharf, had lurched on to its side. 'Now that,' I told Joe, 'was a *real* storm.' Once again, though, the worst – or, from our point of view, the best – had passed through after dark. I wanted some action during daylight hours.

A floating hotel in Pusan harbour capsized by supertyphoon Maemi, September 13, 2003. The strongest storm to hit South Korea in a century, Maemi made 25,000 people homeless and flooded huge areas of productive land.

Hurricane Isabel moving north-west from its birthplace in mid Atlantic, September 9, 2003. Four days later it threatened Florida, before switching direction and making landfall at North Carolina.
NOAA

HURRICANE ISABEL	
Duration	September 6–20, 2003
Highest sustained wind speed	270 kph / 168 mph
Lowest barometric pressure	915 hPa
Damage	US$3.37 billion
Casualties	53
Areas affected	USA (North Carolina, Virginia, West Virginia, Delaware, Maryland, District of Columbia, Pennsylvania, New Jersey, New York, New England), Canada (Ontario)

Hurricane Isabel had started life as a tropical depression out in the Atlantic, practically due east of the West Indies, on exactly the same day Maemi had been born on the other side of the globe. Named at around six in the morning of September 6, Isabel had made her way toward the Bahamas and begun to bulk up big-time. On September 11 she had been upgraded to Category 5, and a couple of days later she had headed west-nor'west to threaten Florida. By this time, Joe and I had finished with Maemi and had made the decision to go stateside and join the party.

On September 16 Isabel altered course again, this time veering north-nor'west to make a beeline for North Carolina. She had lost a little oomph – dropping to Category 3 – but she was still an awesome-looking storm, and states of emergency were declared in North Carolina, Maryland and Virginia ahead of her arrival.

Long flights are the bane of a storm-chaser's existence. The whole way from Tokyo to Los Angeles, we had no way of knowing what was going on with Isabel. The moment we landed at Los Angeles Airport I managed to raise fellow storm-chaser Jim Edds. Jim was already in the little town of Nags Head, which is located alongside one of the beaches in the long sandy islands that comprise North Carolina's Outer Banks. He had booked us a room at a beach-front hotel with a nice view of the water. I told him we'd see him soon for coffee.

We flew to Raleigh and rented a four-wheel-drive. As we passed through the little towns strung along the Outer Banks, we stopped occasionally to film hurricane preparations. Some people had dreamed up jokes for billboards or storm shutters. 'Isabel – Go to hell,' one read. 'Under Water Paint Sold Here,' read another. Still another advertised 'Isabel Party Supplies'. When we finally arrived in Nags Head, Jim was there to greet us. Jim had started out as an underwater photographer, but his love of extreme weather had soon propelled him into the chasing business. Cyclones, he noted, provided plenty of opportunity to film in the wet. I had corresponded with him on email for a couple of years before finally meeting up with him on a tornado chase. This hard-case American had helped me on many of my chasing enterprises, and I returned the favours where I could.

During the night of September 17, Isabel began to make her presence felt. Still on the cusp of Categories 2 and 3, she was a storm to be reckoned with

– and she hadn't wavered from her collision course with our position. The next morning the rain was already driving horizontally. Travelling in convoy, we first drove to the beach and filmed the heavy storm surge battering the pier – which looked to be a terminal case – and then to the strip of affluent holiday homes on the beach front. Jim had a barometer aboard, and reported a sharply dropping glass. It was 11 a.m. By now, Isabel's eye was striking land only a few miles away. Her strength was dissipating, but you wouldn't have known it from where we stood.

As we drove along the road behind the dunes, a few inches of water slid across the tarmac. The sea. The waves were breaching the dunes, finding their way under, around or through the buildings seaward of us, and covering the road. It was a mesmerising sight. We knew that, to be reaching us, the combined swell and storm surge had to be 6 metres, or 20 feet, high. More and more water poured over the road. 'It's a bad business,' I shouted into Joe's camera. 'That's why we're here.'

At one point, doing a piece to camera in front of a 'Welcome to Nags Head' sign, I heard a groan behind me and turned to see one of the buildings that was being pummelled by the waves and wind beginning to come apart. A loose sheet of iron was flapping ominously. To be hit by that going at 100 kilometres an hour would not be good. I jumped back into the car and we beat a retreat. At every turn, though, the sea seemed to be there to meet us. At one stage, excitedly yelling 'Bigger wave! Bigger wave!' I filmed a larger-than-usual flurry of sea foam tumbling toward us.

'Ah … bigger wave moving whole car, Geoff,' Joe said, with just a trace of anxiety.

Jim and I were keen to get shots of swells actually hitting the beach-front houses, so we parked the vehicles where we thought they'd be safe and made a dash on to the balcony of one of the houses. Waves were storming up the beach and bursting among the pilings of the house next door. Jim, dressed in raincoat, shorts and running shoes, his only safety precautions a hard hat and a pair of safety glasses, produced his underwater camera and proposed to go underneath the building to film the surge up close and personal. Well, I was happy to film him doing it.

So with Joe shaking his head and my keeping the camera on him, Jim dashed across the sand to the next building, knelt among the timber piles and

I attempt to measure the wind speed in Nags Head, North Carolina, at the height of Hurricane Isabel, September 18, 2003. Swells, winds and storm surges destroyed piers, houses and businesses; total damage was over US$300 million. JOE MORGAN

calmly pointed his camera at the monstrous, oncoming waves. It was clear the old storm-chasing rule applied – he was either going to get some amazing footage or get killed. And for a moment, if you were the betting type, you'd have put money on the latter. A couple of huge waves pounded up the beach and poured through the timbers. Jim disappeared. 'He's gone!' I said, even as Joe said, 'There he is.' Jim was staggering, but still there. Then another, even larger, wave arrived, sucking him out from under the building and dragging him towards the ocean. Somehow he was able to regain his footing.

As Jim began to make his way around the far side of the building, a giant wave arrived. It exploded through the pilings where he had just been, and roared along both sides of the building. Jim was knocked off his feet and we saw – and filmed – his head going under a tumult of sandy foam. After a moment, he miraculously reappeared. He was being carried by the surge towards the next row of buildings. Some kind of huge air-conditioner or refrigeration unit on a steel frame was rattling around in the water with him. He managed to fend off the heavy machine as the water threatened to drive him against it, or under it. Then the sea carried him into a gap between two buildings, around a corner and out of sight.

A few minutes later he reappeared, clutching his camera and wearing a crazed grin. You could see the adrenalin in his eyes.

'That was extreme,' I said to him.

'Yeah.' He giggled.

During the afternoon we filmed what we could, including a totally destroyed petrol station, but in the evening the police imposed a curfew. We spent a miserable night in our hotel room without power or hot water. Overnight, Isabel moved on to wreak havoc across Virginia and Maryland before being snuffed out in West Virginia.

The next day dawned sparkling. The sea had retreated to its own domain, and was back to its prettiest azure. Placid little waves crumpled on the sand. The ruins of the pier leaned drunkenly out of the water. The road which had run along behind the dunes was no more, the tar seal stripped away and replaced by sand. Some buildings had been obliterated, others picked up off their piles and dumped some distance away. We found one of the owners gazing sadly through the window of his wrecked beach house. He agreed to let me rescue the television set inside for the benefit of the camera. He even contributed a line to the story. As I staggered across the sloping floor, clasping the television set in my arms, he called out, 'Can you keep an eye out for the remote control?'

OPPOSITE

At the height of Hurricane Isabel, a storm surge undermines houses and washes away a road in Nags Head, North Carolina. My fellow storm-chaser Jim Edds was knocked off his feet and nearly killed while under this house trying to film oncoming waves. September 18, 2003.

The Big Chill – Oymyakon, Siberia

The **Dangerman** *series presented me with the opportunity to tick off a few things on my all-time wish list, and travelling on the Trans-Siberian railway was right up there. The trouble was how to pitch it to Pioneer.* I tried out: 'Come with me as I travel on the world's most extreme railway,' but I could tell that didn't really capture the producers' imaginations. Such an episode didn't stack up against filming in the middle of a Category 5 hurricane and abseiling into the crater of an erupting volcano.

After some hard thinking I hit upon the idea of using the train ride to get to the coldest permanently inhabited place on Earth, Oymyakon in eastern Siberia, where the average temperature in winter is -60°C (-76°F). Pioneer bought it, so in the middle of January my colleagues Mark Whetu and Rachael Wilson

The Trans-Siberian Railroad, completed in 1916, opened up Earth's coldest inhabited region to trade, exploitation of natural resources – and occasional adventurers! The entire journey of 9441 kilometres (5867 miles) from Moscow to Nakhodka takes 8 days, and includes a compulsory overnight stay at Khabarovsk, a former military outpost, now home to half a million people.

OPPOSITE
Life on the streets of Yakutsk at -40°C (-40°F). The capital of Sakha Republic in Russia's eastern Siberia, the three-centuries-old city is built on permafrost 140 metres (450 feet) deep.

and I headed for Moscow. Once there, we picked up a guide, Rob Walker, an American fluent in Russian, and headed off on the world's greatest train journey, 4500 kilometres across Russia and Siberia, crossing eight time zones.

The terrain was spectacular but it was hard to get good photos out the window. After we tried and failed to bribe the train driver to let us ride with him in the cab, we were allowed to ride in the baggage car, which had a view out of the rear of the train. To get to the best position we had to clamber over what we thought were frozen cattle carcasses wrapped in sacks. The temperature in the car was -20°C (-4°F) so frozen food happily stayed frozen, and the frozen items turned out to be convenient places on which to sit while we filmed. Later, through our interpreter, we found out with a shock that the 'carcasses' were in fact the frozen dead bodies of soldiers being returned home from the war in Chechnya.

After all that distance the train arrived precisely on time, and from the terminus we flew 1500 kilometres to Yakutsk, the capital of Siberia's Sakha Republic, and then drove 1000 kilometres to our destination. Oymyakon has been known to drop to -70°C (-94°F) in winter, but it was a comparatively balmy -45°C (-49°F) when we arrived. A local tour operator had arranged for us to stay with the family of a farmer as we prepared to carry out the stunt we had devised.

ABOVE AND RIGHT
Living in a deep freeze: Yakutsk, Siberia. Amazingly, in summer the temperature can climb to 30°C (86°F).

TOP LEFT
On the road from Yakutsk to Oymyakon. The road was built by Gulag prisoners in the Stalinist era from the 1920s to 1950s.

TOP RIGHT
A classic stunt: the cup of boiling water I throw in the air in Oymyakon freezes and falls to earth as powdery ice crystals.

LEFT
The village of Tomtor, 40 kilometres from Oymyakon, where the world's coldest temperature outside Antarctica, -71.2°C (-96.16°F), was recorded in 1926. The temperature was -47°C (-53°F) when this photo was taken.

Mark and I were going to sleep outdoors overnight in a tent, perhaps the only people ever to have done so in Oymyakon. In a lot of ways, the experience reminded me of camping out in the backyard of my family home as a child, except that in those days we got by with pyjamas and a cheap sleeping bag. For Camp Siberia, Mark and I were trussed up like Teletubbies in thermal suits, and jammed into super-warm sleeping bags. To add spice for viewers, before I settled down for the night I demonstrated one of the curious effects you encounter at temperatures as low as this, carrying a cup of boiling water outside and throwing it into the air. The water instantly froze and fell to earth as a shower of powdery ice crystals.

Oymyakon was a unique experience, but for all we talked up the adventure on the screen the most dangerous part was when I filmed Rachael too early in the morning after our first night on the train, before she'd had a chance to have a cup of coffee.

The Fiery Pit – Mount Marum, Vanuatu Erta Ale, Ethiopia

I had long dreamed of getting up close to boiling lava in a volcano. There seemed to be two ways this could be done. Either some sort of vehicle could be driven down into a suitable crater, and parked as close as possible to the margins of the molten material. Or a cable could be strung across a crater, and a pulley system used to lower the cameraman (me) down to the crater floor. Either way, I would wear a heat suit, and film as much as I could from close range. Who knew how useful the footage itself would be, but the attempts to get it ought to make breathtaking television.

One of two boiling lava lakes in Mount Marum, one of Vanuatu's many active volcanoes, March 17, 2004.

I came up with two plans. Plan A was to drive a bulldozer down into a crater, requiring a volcano with gently sloping walls. Plan B, to abseil into the crater, required a volcano with sheer walls. I figured the volcanoes in Vanuatu had enough variety to cater for both options, so I got in touch with Harold

Tavner, the pilot who had flown me to Tikopia during Cyclone Zoë, and asked him to source me a bulldozer. Naturally, he asked what I wanted it for. I told him I wanted to drive it into a volcano. Harold said he would see what he could do.

On December 28, 2003 Rachael and I flew to Port Vila. That night we had dinner with Harold and convinced him we were serious about the bulldozer stunt. Over the meal we cooked up his lines for the show, and the next day, when we drove around to Harold's place, he pretended to be surprised to see us.

'There's gotta be a cyclone or an earthquake or a tidal wave to bring you around here,' he said cheerfully. 'What is it this time?'

'A volcano,' I told him. 'I'm looking for a bulldozer.'

'A volcano. What's a bulldozer got to do with a volcano?'

'Well, I want to drive a bulldozer into the crater of a volcano,' I told him casually.

'Into what?' he said. 'Into *what*?'

He gazed off into the distance.

'Into,' he mused, as though his mind was already on the job. When it comes to acting, Harold is a natural.

Since we'd spoken he had managed to scare up rumours of two potentially suitable bulldozers. 'Suitable' in this context meant 'in working order' and, just as important, handy to an active volcanic vent. Both machines were on the island of Tanna, which sounded good: there were two major active volcanoes on Tanna, including Yasur, into which I had previously abseiled in 2000. I knew it would be eminently suitable for this escapade.

Harold flew us out to Tanna in his trusty Cessna, a couple of hours over astoundingly beautiful blue ocean studded with turquoise and green atolls, and once on the ground we rattled over awful roads to the Public Works Department compound, which sounded far grander than it proved to be. A dilapidated wire fence enclosed a tin shed, around which lay a collection of ruined machinery, including a huckery old bulldozer. By the look of it, the machine had been exposed to the elements for some time. And if rust never sleeps in temperate zones, it's on amphetamines in the tropics: the whole thing was an orangey-brown, greasy mess. I didn't really need Harold's expert opinion on its mechanicals to know that this thing wasn't going to be much use.

The other machine belonged to an evangelical mission up in the bush. Word had it that, although this machine was broken-down, it was 'only slightly broken-down', making it a better prospect than the severely broken-down Public Works machine. We thought it was worth a look. We raced up to its supposed location, insofar as it is possible to race over roads like those, but arrived in the dark. Here, from an American missionary who appeared to be in charge of the place, we learned the bad news: the bulldozer had been abandoned precisely where it had broken down, 16 kilometres away, at the end of the track it had been gouging through the jungle.

We celebrated New Year's Eve back at base on Tanna. Then, on the first day of 2004, while the rest of the world slept off its hangover, we headed back up to the mission and set off on quad bikes to check out the bulldozer. The further we went down the dirt track, crowded on both sides with jungle and the occasional sugar plantation, the further we were getting from the volcano. Coaxing the machine to the volcano was clearly going to be an epic in its own right, let alone what came next.

The bulldozer turned out to be a little Case, probably equivalent in size to a Caterpillar D1. It was sitting forlornly in the clearing where its mishap had occurred. Despite its predicament, it had all the appearances of a well cared-for machine: it was covered with a tarpaulin, and there was a milk-powder tin over the exhaust to prevent water getting in. Harold was sure he could get it going. I'm not sure how precisely he explained our project to the missionary, but the man, who was small and bearded, was happy to hire it to us. We shook hands with him, and with every other member of the curious group which had gathered to watch the negotiations. We had a deal.

Next, we headed for Yasur volcano itself. We walked up the cinder slopes, scouting a route to the crater's edge. Harold, who had far more experience driving bulldozers than I did, was confident we could head straight up the lower slopes, and then sidle across the side of the mountain as it became steeper. He was less sure how we could get from the crater rim to the ledge overlooking the lava lakes. We stood peering down. It looked to me as though it would be possible to drive the bulldozer down and park it right at the brink.

Harold shook his head. 'Nah,' he said.

In front of the camera, I was upbeat. 'This will be my most dangerous undertaking ever,' I said portentously. I flashed the *Dangerman* grin.

Against the possibility that Plan A wouldn't work, we also reconnoitred for a Plan B. Harold flew us to Ambrym in the north of the Vanuatu group, where we overflew Mount Marum. Marum was ideal for my purposes – a relatively compact circular crater with nice sheer walls, and the likelihood of good holding around the rim for the cable anchors. And down in the engine room there was a beautiful, heaving cauldron of orange lava. Meanwhile, though, we proceeded as though Plan A were full steam ahead. Rachael and I returned to Auckland, where I took the test for a tracked vehicle endorsement to my heavy trade driver's licence. This wasn't strictly necessary – I doubt whether anyone on Vanuatu gave a tinker's whether I had a proper licence or not – but Pioneer was keen, for insurance purposes, for me to hold the right papers. There was also something fantastic about making sure I had a licence to drive a bulldozer into a volcano.

Although I was to take the test in a highly controlled environment in Auckland, when we discovered there was nothing in the regulations which stipulated what kind of tracked vehicle you had to use we sensed the opportunity for an amusing *Dangerman* stunt. Joe Morgan knew a man who owned an ex-Australian Army tank and was willing to let us use it. We arranged for Peter Best, a real driving instructor, to pretend to put me through the driving test. We filmed Best fiddling with a clipboard while I trundled the tank over a low hill to the watercourse where I was to be put through my paces.

'Holy hell!' Best yelped authentically when he saw the tank, 'What is this?'

He recovered sufficiently to agree that, yes, the principle was exactly the same. And yes indeed, there was nothing to say I couldn't do the test in a tank.

To add televisual interest, I took the test for a 'T' endorsement to my driver's licence, which would allow me to drive tracked vehicles, in an ex-Australian Army tank.

I duly performed the series of tasks – reversing, cornering, rolling over difficult terrain – to his satisfaction, and he ticked the necessary boxes on the form. For the purposes of the show, this meant I could now legally have some fun in the tank and I spent a happy half-hour driving over and crushing a car wreck, and bursting through an old shed thoughtfully put there for the purpose.

A few days later, we received bad news from Vanuatu. Harold had been unable to get permission from all the appropriate chiefs for me to drive a bulldozer into their volcano. The sticking point, as it often is, was the price. Yasur is the only real asset on Tanna as the volcano draws a good number of tourists every year. The road that led to the mountain was in poor condition, having been repeatedly washed out by flooding, and we had offered to use the bulldozer to repair it. This would have been of immense value to the locals, but they continued to insist on a hefty payment. When we couldn't convince them, it was back to Plan B – abseiling into Mount Marum.

I got in touch with Mark Whetu. I had first met Mark, a fearless adventure guide, a few years back when covering the Southern Traverse, a mountain-running race in New Zealand's Southern Alps. He and I had been caught by a storm, and had had to camp out for a couple of days. When the weather showed no sign of improving, we had decided to walk out. Along the way, I fell over a small cliff in the fog and Mark had to rescue me. I was shaken but unhurt. I swore him to secrecy, but somehow word got out to my colleagues at TV3. My little mishap was on the news and I was not happy. Still, when you meet an adventure technician of Mark's calibre, you can forgive him the odd indiscretion. He had joined Rachael Wilson and me to make *Storm Shooters*, and when I signed up to do *Dangerman* I had lost no time getting him on board.

Mark devised a system which we decided to try out over a wide gorge called the Rakaia, a short drive from my hometown of Christchurch. The first step was to use a jet boat to take a steel cable across the river. The idea then was to stretch the cable taut, so I could be hauled out along it on a pulley until I was ready to lower away. Then I would lock the pulleys on to the cable, drop a climbing rope and abseil down this with my gear. Simple! The trouble was, we had got hold of sophisticated abseiling gear for the purpose and I couldn't make head or tail of it. I spent a long time dangling over the river trying to figure it out while Rachael rolled the camera and laughed. I wasn't too happy when that segment made the final cut.

With everything now ready Rachael, Mark and I returned in late March to Vanuatu, where we were joined by Phil Cotter, a local helicopter pilot, Harold Tavner and Daniel Lewis from Australia. I had first met Daniel after a storm-chasing jaunt to Western Australia. I was getting off a plane in Perth when one of my fellow passengers said, 'Are you Geoff Mackley? I've watched all your shows.' Daniel, a 26-year-old air-conditioning technician from Perth, was also an aspiring adventure nut. We kept in touch and finally I had had the chance to offer him a place on an expedition.

After we lost a day which I spent vomiting, bugs being always a hazard on tropical trips, we made our way by Cessna and helicopter to Ambrym Island, and prepared to reconnoitre the volcano. Mark, Phil and Daniel took the helicopter up, but soon returned. Although weather conditions were good at the little grass-strip airport, it was a different story up the mountain. Because of its height and all the heat and vapour it generates, Mount Marum creates its own climate. On this day the summit was totally closed in. This meant another day lost, and an uncomfortable night spent camping in the airport terminal – or the plywood shed which bears that name.

The following day, March 17, there was a break in the weather and we seized the chance to get our gear – which included more than a kilometre of assorted steel cables and rope, about a tonne of drinking water, and such specialised equipment as gas masks – and ourselves to a suitable landing spot near the summit of the volcano. It was an awe-inspiring landscape to fly over, a huge swathe of jungle scorched and charred by the mountain's eruptions giving way to the barren moonscape of the cone itself. We flew four trips to shuttle our equipment, and then Phil dropped us off near the summit, where

Our camp on the summit of Mount Marum, Vanuatu. It was desperately hot, and just a short walk away from two huge, boiling lava lakes.

Mount Marum spurts lava high in the air as we make plans to abseil across the boiling lake.

we spent a few hours toiling in the desperate heat, shifting gear and setting up camp near the crater's edge. A very short walk from where we pitched our tents, we could look over the rim and see the two huge, turbulent lava lakes seething and boiling. They were nearly 300 metres down, a lot further than we'd bargained for.

As the camera rolled and I started explaining the task, a huge spurt of lava shot from one of the lakes. In the film, the material seems to hang in the air for an age, the only true gauge in such an immense landscape of the scale of the eruption.

'I reckon this ups the ante on the danger stakes, Geoff,' Rachael chirps from behind the camera.

'Oh, it's a little bit dangerous,' I say, with a pensive expression on my face. 'A little bit dangerous,' I repeat, scratching my nose. What you can't see are the butterflies which were churning about in my guts, easily as vigorously as the molten rock hundreds of feet below.

We spent the rest of the day mapping out the task. It was considerably bigger than we'd first thought. The most promising route for the cable was from a point not far from our campsite, across to the summit of a mound on the opposite edge of the crater. It was a long way across, and a long, long way down to the point where I wanted to go. Mark was concerned that, because the crater walls weren't vertical, there'd be plenty of scope for chafing and damage to the ropes before I even got airborne along the cable. And it looked as though I'd struggle to make it all the way down, although I was certainly going to have a damn good crack at it.

By the time we thought we had it all worked out, and I was explaining the plan to the camera, things had started to go wrong. This is clear in the film: my face is barely visible for water on the lens, and the molten lava is a dim orange circle in the mist behind me. The scene could well be a barbecue filmed through a foggy window. The rain was pelting down, and the wind was beginning to come up. Wind and rain on a mountain aren't ever much fun, but on the top of an extremely active volcano I was struggling to find words to express the unpleasantness. Acrid clouds of toxic gas – principally hydrogen sulphide, sulphur monoxide and sulphur dioxide – had been blowing through intermittently all day. Now they were being whipped out of the crater by high winds and streamed directly over us. Our eyes, noses and throats were all burning.

MOUNT MARUM	
Location	Vanuatu
Type	shield
Height	1334 metres / 4377 feet
Last eruption	2006 (ongoing)

By midnight, it was clear this was no passing squall. We were in for a major storm. Knowing there would be a highly acidic environment near the crater, we hadn't splashed out on expensive gear as it would inevitably be damaged. This proved to be false economy. Our cheap tents were flogging themselves to bits, and had filled up with sulphurous fumes. We were cold, soaked through and gasping for breath.

'As a holiday destination you've brought us to, Geoff,' Rachael said, 'this sucks.'

The storm continued relentlessly all night and into the next day, and the next, and the next.

By day four, the wind was exceeding 80 kilometres an hour. There was no possibility of pursuing our plan to rig the volcano. We spent every hour, day and night, holding up the tents, trying to mend broken poles with gaffer tape and weighing everything down with water containers and Pelican cases. At some point the wind began to nudge cyclone range: I estimated that some of the gusts were barrelling through at 100 kilometres an hour. I yelled, 'It's absolute mayhem here!' to camera inside our wrecked tent. It seemed ironic that just when I had decided to have a spell from chasing storms, a storm had stalked me. In idle moments, trying to hold the tent together, I wondered what its name was.

Our camp at the summit of Mount Marum, decimated by a storm. I usually love storms, but this one was definitely an exception.

We repaired the tents as best we could, but overnight the wind would rise again and flatten them. To make things worse – yes, this was possible – my little Honda generator overturned, petrol mixed with the water on the floor of the tent and its fumes with the sulphurous muck we were already breathing.

We had an aviation radio with us to communicate with Phil but by March 22 – day six of the expedition and day five of the storm – the batteries had died. Fortunately, we managed to raise him by calling the airport at Port Vila on my satellite phone, and getting staff there to radio him. He had been on standby to come and extract us for nearly 48 hours, but the weather hadn't permitted it. He was still in bed, the forecast suggesting staying in bed was a more sensible way to spend the day than flying a helicopter to the summit of a volcano in a hurricane.

Things were looking black. Daniel was all for walking out. Mark and I were against it. I had made the trek before and knew what was involved. It could be done, but only if you were desperate. I didn't consider we were yet in a totally desperate situation. While conditions were far from pleasant, they were survivable. And ironically the climate at Marum's summit was nicer than that in the hot malarial swamps lower down. Although I wasn't afraid, I was annoyed I had got my colleagues into this mess. All our gear was wrecked. Our tents were shredded. What tins of food remained were orange with rust. The rain had combined with sulphuric fumes to create sulphuric and sulphonic acids – the constituents of acid rain – and this was playing havoc with both our gear and our bodies. Any exposed skin was burning like crazy, and the skin on the palms of our hands was hanging off in white blisters. Our eyes and noses were stinging horribly.

Mark and Rachael's tent was now unusable. If we were going to spend another night up here, they needed shelter. We decided to try to scratch a cave in a low bluff nearby. We made good progress – the ground we were digging was brittle and crumbly – and that was the problem. The cave collapsed, almost burying Mark and Rachael and actually interring some of their gear.

Among the rigging equipment we had left at the summit was a set of huge, heavy canvas bags we had brought to fill with gravel and use as anchors. Daniel and I decided we would head up and get them so we could use them as bivvy bags – makeshift, one-person tents. Just as we reached the summit at about three in the afternoon, we heard the throb of rotors. Phil's helicopter appeared beneath the lowering cloud, hunted along the ridge for a stretch, and then, to our wild delight, set down. Daniel and I raced back down to the campsite, where a stand-up argument ensued over who should leave the mountain first. I thought Daniel and Rachael should go while I got our gear together. Rachael, however, refused. We were there to do a job, she said, and someone should be filming the evacuation. Mark thought Daniel and I should leave first. Phil, meanwhile, was fuming. He couldn't leave his machine idling in the acid rain forever. In the end, Daniel and I were the chosen ones. As we rose into the air, Rachael, Mark and the ruins of our campsite were swallowed by the gloom.

After five days on top of Mount Marum, we called it quits. Not all adventures work out the way you want; this one had turned into a living nightmare.

For a while it seemed as though the break in the weather which had allowed Phil to get to us had been a one-off, and Rachael and Mark would remain stuck up in that hell-hole, with the forecast predicting worse to come. Luckily, though, Phil got another sniff of a break. That was all he needed. The conditions were worse than marginal: not many helicopter pilots could have pulled off such a rescue, even if they'd dared to try. Phil described the mission as 'scary'. Coming from him, we knew it had to have been bad.

Once we were all safely down, we wanted to buy Phil a beer. The trouble was that our wallets were on the summit of the volcano, along with our passports, our clothes and even the computer that held all our digital photographs. We were like refugees. Joy Wu from the Kaiviti Village Hotel kindly came to the rescue, taking us into town to buy some basics, and lending us money to tide us over until we could fly a salvage mission to the campsite. In nearly 20 years of adventuring, I had never lost a single item of gear. Now it looked as though I had lost almost everything I'd taken on to Ambrym, except my determination to get closer to a lava lake than ever before. We drank a toast to the worst expedition ever. By the time three or four beers had slid down, Mark and I were deep in discussion about finding another, more suitable volcano.

For my second attempt, Erta Ale appealed on several levels. It was a classical shield volcano with broad, gentle slopes, which would make it relatively easy to climb, and it had more or less permanent lava lakes at the bottom of a couple of sheer-sided craters, which would make it perfect for rigging.

Better yet, it was in an extreme location, right in the middle of northern Ethiopia's Afar Depression, which is well below sea level, and not only one of the lowest places on Earth but also one of the hottest and, after Antarctica, the driest. Ever since I had first seen a picture of the mountain I'd been keen to go there. Best of all, there was no chance of a cyclone coming along: the region had last seen rainfall when Adam was chucking away his apple core.

Back in New Zealand, Mark and I, while waiting for the big day, refined the cable and pulley system. We had started to realise it was asking a lot of the person making the descent – in this case me – to haul himself up with the rope-climbing device known as a jumar. A jumar is basically a clamp that slides up a rope and grips when weight is applied. It can slide only up, not down. With two of these you can virtually 'walk' up a rope. However this would be nearly impossible when wearing a heatproof suit at 50°C (122°F). And it was asking a lot of the crew at the top to haul up me and all my gear. We decided the retrieval could be done with a little electric winch, the kind used to haul boats out of the water and on to trailers. We added one of these, and a truck battery to power it, to our kit.

On February 20, 2004 Rachael, Mark and I flew to the Ethiopian capital of Addis Ababa, and once there set about trying to track down the things we needed to travel 1000 kilometres through the wilds of the country to Erta Ale – namely three four-wheel-drive vehicles, eight camels, some armed guards and a tonne or so of water. This was my first time in Africa, and Addis Ababa was an eye-opener. It looked as though the Middle Ages had collided with the twenty-first century. Ethiopians are a striking race, tall and proud, with long noses and a solemn cast to their features, and they wear an incredible variety of coloured clothing. On the streets, beaten-up cars jostled with creaking wooden donkey carts and pedestrians carrying huge loads on their heads. I was told you could buy just about anything in the market and I didn't have much reason to doubt it.

Although Erta Ale is one of the most remote spots on Earth, these days there's something resembling a beaten path there, with several expeditions visiting every year to study the volcano. The success or failure of these expeditions usually reflects the level of organisation. I had taken the advice of a party of French volcanologists who had made the trip the year before, and who had been recommended by Jacques Durieux, a famed French volcanologist who is an expert on Erta Ale and has been there more than anyone else in the field. We came away having secured the services of a guide, Ashrad, several escorts and three vehicles. Through a guiding company, we had also arranged for the Ethiopian Air Force to bring our water by helicopter to our campsite on the summit of the volcano.

ERTA ALE	
Location	Afar Triangle, Ethiopia
Type	shield
Height	613 metres / 2011 feet
Last eruption	2006 (ongoing)

The next day we set off through the kind of landscape I'd only ever seen on television – tall grasses, fever trees and wild animals of every description. We spent the first night camping next to a river in the Awash National Park, none too thrilled about the crocodiles drifting about in the beige water close by. Rachael and Mark chose the two tents farthest away from the river, as if crocodiles could be relied on to go to the first tent they found. As it turned out, though, none of us were bothered, and our escorts weren't called on to use the evil-looking curved knives they carried.

The next day we bumped and ground our way north for another eight hours, before stopping in the compound of what we were told was an abandoned school. It wasn't abandoned. The next morning, a steady stream of skinny kids clutching school books materialised from the desert. We were the centre of attention, and I entertained them with a trick which never fails, taking their pictures with the digital stills camera and showing them the image on the viewfinder.

Three or four hours on, we reached Hadar. This should have been a huge thrill for me. Hadar is the site where in 1974 a young American paeleoanthropologist called Don Johanson found the fossilised remains of man's oldest, most complete human ancestor, *Australopithecus afarensis,* which was to gain the nickname Lucy. The following year Johanson and his team also found the fossilised remains of about 13 more human ancestors, apparently living together, who were subsequently dubbed The First Family. Hadar was, therefore, the birthplace of humanity, the spot where it all began, but it was pretty inauspicious to look at, barren and lifeless. I longed to go exploring but it was not to be. A combination of the heat, four hours in a dusty lurching Land Cruiser, and something I had eaten had made me ill. I lay fitfully in the shade while Mark and Rachael poked about on the shingle slopes.

Two or three hours beyond Hadar, the main road ended in one of the world's most isolated human settlements, nestling alongside the reason for its existence, Lake Afrera, from which salt is extracted. If anywhere had the look and feel of another planet this was it, the hottest, driest and, at 100 metres below sea level, lowest place on Earth. It was 43°C (109°F) and the town had the look of a movie-set from hell, with hundreds of makeshift shanties made from anything and everything. We called in at the police station, having been advised that beyond this point we were in bandit country.

A party of Italian explorers had been kidnapped a few years before and held for three weeks. It was clear that to impoverished people our expedition could represent a convoy of unimaginable wealth. The police told us we had better spend the night in the station itself if we couldn't scare up some guards. In the end, we managed to recruit six likely looking men equipped with even likelier looking assault rifles.

We also found hot springs, our first evidence of the volcanic belt of which Erta Ale is a part, and had our first decent wash for a week.

A 'dust devil' makes its way across the barren landscape on the road to Lake Afrera – one of the earth's most remote and low-lying places.

Erta Ale is one of the loneliest places on Earth but a few intrepid volcanologists find their way there – despite the difficulties encountered with transport, desert, dust storms and warring villages.

The next day, it was still a six-hour drive from the end of the official road at Lake Afrera to the point where we'd probably have to rely on camels for transport. We had gone only a couple of kilometres when one of our vehicles became stuck in soft sand. As we tried to dig it out another went in up to its axles. This was not a good situation. There was no point yelling for help: we were on our own. Finally, after copious digging, shoving, revving and swearing, we managed to extract the vehicles and carry on.

We now found ourselves trundling over a landscape which looked the way I imagine the surface of Mars. This was the Danakil Desert, barren, bleak and godforsaken. Rocks, as far as the eye could see, were strewn over sand and shimmering in the fierce heat-induced mirage. The only things moving were dust devils – mini-tornadoes that whipped up the sand into little funnels, which then drifted across the plain and disappeared.

With every kilometre we could see Erta Ale, and the signs of nearly 100 years of continuous violent activity, more clearly. Even to have got this far was something only a handful of Westerners had achieved. To go any further, we had to have the trust and assistance of the local people. And yes, there were local people. At the foot of the volcano were two villages, each consisting of crude shelters of dried earth domed over wooden frames. As for livestock, there were a few scrawny goats and, of course, camels. These people, the Afar, were largely nomadic herders who travelled long distances in search of feed for their animals.

No sooner had we arranged to hire eight camels from Dodom camp, the first village we reached, than word got back to the other one and a dele-

Hiring camels for the last leg of our journey got us involved in power struggles between two neighbouring villages.

gation arrived to insist we hire *their* camels. It was hard to say who held the superior bargaining position – both sides were heavily armed with AK47s – so we left them to hammer it out between them. A couple of hours went by, with no resolution. We made our way to the rival village in the hope this might produce results. Who knows what compromise was reached, but in the evening a line of ten camels appeared, saddled up, complete with drivers and ready to go. We wanted only eight, but if paying for another two preserved the local political balance it was, we decided, money well spent.

We were anxious to get moving as soon after dark as possible: it was too hot to travel during the day. As soon as we could get organised, strapping all our gear to the camels and settling ourselves up on the saddles – a first for all of us – the big maudlin beasts were coaxed groaning on to their feet and we were on our way. Five hours and 15 kilometres later we were lost. The landscape was featureless, apart from the volcano, which we couldn't make out in the dark. I'd expected to be able to see a glow from the crater, but the horizon was pitch-black. We set up camp and the next day, before the worst of the heat, covered the rest of the distance – around 25 kilometres.

The following day we started the climb up the volcano. As volcano climbs go, if it weren't for the heat the gentle slopes of Erta Ale would be among the easiest in the world. Although you make quite a lot of height from the plain to the summit, when you're standing on the top you're only 690 metres above sea level, because Erta Ale begins from land well below sea level. Just before we reached the crater, our porters sat down and refused to go any further unless we paid them more. I was exasperated but not surprised. Everywhere you go in the world, local people will impose a 'crater tax' when they guide you up a volcano. And we'd come too far now to decline.

As we were setting up camp, to our profound relief we heard the noise of a helicopter, and the air force whirlybird that was bringing our water appeared. This improved our chances of survival. No sooner had the water been off-loaded, however, than scuffles broke out among the camel drivers, who had neglected to bring their own supplies. It was hard to stand by in the searing heat and watch our precious water being slopped over the ground by men having a tug of war with a jerrycan, but given that each had an assault rifle slung over his shoulder, standing by seemed the prudent thing to do.

Once our camp was in place, guaranteeing us a refuge from the oppressive heat, we took a look into the crater. I was dismayed. You could see the lava lakes, but they were completely crusted over, dull, grey and lifeless-looking, apart from the odd pinprick of orange glowing through at the top of one of three 10-metre-high hornitos. It was a long way to come to see ... nothing.

Still, we had a job to do, and do it we would. Mark and I surveyed the crater rim for suitable spots to anchor the cable. Mark was pessimistic at first. The material around the rim was volcanic slag, or scoria, incredibly light and porous. Often it would break away in large chunks. Mark used his portable drill with an auger bit to sink test holes. Time after time, the bit sank into the rock as easily as a nail into timber. Things were looking grim.

Finally, persistence paid off. Mark located a suitable area on each side of the crater where he was able to drill holes in nice solid rock. He then epoxied bolts into the holes, and attached the anchors, a V-shape of chain, to the bolts. I was a bit dubious about staking my chances of avoiding an uncontrolled plunge into the crater of an active volcano on a tube of glue, but Mark had faith. 'I reckon that'll just about hold Ethiopia together,' he said with satisfaction, as he tapped the last bolt home.

Next, we walked a guide rope around the rim of the volcano and rigged it to the pulley system which would conduct the steel cable over the void. As we were doing this, around 5 p.m. local time, the volcano suddenly came to life. There was a kind of quiet snort from the lava lake, amplified by the crater. One of the hornitos had broken open and was emitting a small but vigorous fountain of highly fluid lava. Other cracks were also appearing, threads of vivid orange, all around it. These widened, and whole sections of the crust shifted, tilted and vanished into what quickly became a seething tumult of brilliant lava. Small spurts of the lava were tossed in the air.

Rachael told the camera this was the first time she'd seen lava blown. 'I can't take my eyes off it,' she said. I had seen the phenomenon before, especially when Mount Etna had blown its top in 1999, but it's a sight that never loses its fascination. And as I was planning to view the show from close to spitting distance, it took on a whole new and personal aspect. I wanted to see the show all right, but I sure as hell didn't want to get fried.

At night Ethiopian volcano Erta Ale showed its true nature with a dazzling display of boiling lava.

Finally we got the cable across and anchored, and Mark set to work rigging the pulleys, harness and tether I'd use to move along the wire and then abseil to the ridge. We were keen to have everything up and running that evening, so we could test it all and do the business the next day. As night fell, Mark announced it was ready to go. I'm used to heights, but there's always a swoop in your guts when you contemplate suspending yourself over a void. When the void happens to be the crater of an active volcano, that swoop is a little more pronounced then usual. I trusted Mark implicitly, so I had complete faith in the rig, the individual bits of gear, even the superglue. But this was a pretty mad stunt.

The next morning the lava lake had crusted over again. We couldn't let that worry us. While the air was as cool as it was likely to be that day, I put on a climbing helmet, and Mark buckled me into the harness and snapped on the tethers. I was ready to roll. I eased gingerly into position on the crumbly brink of the crater and, with Mark directing, I eased my weight into the harness and on to the cable. Suddenly I was clear of the crater wall and dangling in space. I let myself glide along the cable a few metres until there I was, suspended over the crater of Erta Ale, 90 metres above the ledge below and 120 above the lava itself.

'This,' I said to myself, 'is the best view in the house.'

The scale defied normal powers of perception. I felt strangely small. Occasionally, wafts of heat from the lava bubbling away below engulfed me, as though the door had been opened to a blast furnace.

When I signalled I was ready, Mark hauled me back along the cable and I scrambled up on to terra firma. I was drenched in sweat, but exhilarated. The cable was going to hold, so now there wasn't anything holding me back. Except, perhaps, fear – but when had that ever held me back? We played around on the cable, each of us taking turns going out and having our photos taken, until the temperature became too hot to do anything and we slumped in the shade. That evening, as though on a timetable, the lava lake awoke again, and by 6 p.m. the whole expanse was boiling and seething.

In the comparative cool of Monday morning, Mark double-checked the rig while I fought my way into my reflective heat suit, actually an airport firefighter's suit designed to protect the wearer from temperatures up to 1000°C (1832°F). By the time I was in it and had assembled all my other gear, it felt as though it were 1000 degrees *inside* the suit. I was burning up, and I hadn't even begun to enter the volcano yet.

I clipped on to the cable, and after a quick run-through of procedures – we'd been over them dozens of times already but one more couldn't hurt – it was time. I allowed myself to glide out to the middle of the cable, with Mark paying out the safety rope as I went. The heat was intense. By the time I reached the point at which I was to make my descent, I was wondering whether I'd be able to stand it.

When I reached up to lock the pulleys in position, which was essential if they were to act as anchors for the rope which would retrieve me from the crater floor, I found I couldn't reach them. They were half a metre too far above my head. I tried hauling myself up with one arm and operating the locks with the other but it was hopeless, and with each attempt I grew hotter and weaker. It would have to be fixed. I yelled to Mark to reel me in.

The plan at first was to adjust the harness while I waited at the edge of the crater, but I began experiencing that terrible panic which goes with heat exhaustion. I just had to get out of the suit. Mark, who hates messing around, was growing impatient, but when he saw how distressed I was becoming he was concern itself. I stripped off the suit as quickly as I could. The rest of the crew offered to tip water over me but there was no need: my clothes were already drenched in about a gallon of my own perspiration.

Abseiling over the crater of Ethiopia's Erta Ale volcano on a high wire, March 2, 2004. The seemingly benign-looking lava lake conceals a seething tumult below the crust. The mountain has been erupting continuously since 1967.

While Mark got the harness ready I did my best to recover and rehydrate. At the same time, I decided the only way we were going to achieve the stunt was if I carried the suit with me down to the crater floor and put it on there. This would introduce another element of risk as I would be left exposed to any escalation of activity which took place while I was descending, but it was the only way I could imagine surviving the suit.

This time, things went smoothly. I cruised to the middle of the cable and then locked the pulleys without difficulty. I unclipped the canvas bag containing the climbing rope that was fastened to the overhead pulley, and let it drop. It fell for what seemed like an awfully long time then smacked on to the rock below. I clipped into the abseil device on the rope, took up the tension, and then unclipped from the tether to the cable.

'Clear!' I yelled. Then: 'Descending!'

Down I went, 90 metres into the fiery pit. It took only a couple of minutes, and I was standing on the floor of the crater. I laid out my gear, had a quick swig of my drink, and then began setting up the camera to film myself putting on the heat suit. I could feel the intense heat striking up from the rocks through the soles of my heat-resistant boots. In the smooth sculpted rock around me there were pores and fissures, from which gas and gusts of intense heat were issuing. I was clearly standing on a crust over the molten magma – a thicker version of the rock crust we had seen liquefy the day before. At one point, I put a carabiner on the ground. When I picked it up a few seconds later, it scorched my hand.

I could already feel the beginnings of heat distress creeping up. I struggled into the suit and then I was ready to walk to the edge of the lava lake. When you're under this kind of physiological stress, you start monitoring yourself for signs of unsound decision-making, anything which might indicate your mind is cutting corners to try and lessen the strain on your system. Everything seemed to be in working order, but I knew I had only a few minutes. I took another swallow of my drink, dropped the heat helmet over my head, and began to walk toward the point where Earth finished and hell began.

If this moment ever came, I had long promised myself, I'd mark it with a few suitable words.

'That's one small step for a man,' I said on tape, 'but one giant leap for mankind.'

And if you think the spacesuit and moonscape had gone to my head and

I was getting ideas above my station, consider this: 12 people have trodden the moon's surface. I probably wasn't the first person to have trodden Erta Ale's crater floor, but fewer people had walked where I was walking than had kicked about on the moon.

I walked as close to the edge as I dared and filmed, but there wasn't a lot to see. The crust was solid-looking and the only sign of activity was faint de-gassing from a couple of small, otherwise inert hornitos in the middle. Nevertheless, the heat quickly became unbelievable. A couple of minutes, I told myself. Any more and I'm pushing the envelope too far.

All too soon it was time to go. I retreated to my gear and got out of the suit. My head was pounding and I was dizzy with the onset of dehydration. I got everything together as quickly as I could and radioed to Mark to begin the retrieval process. I hooked my boot into the loop at the end of the rope, clipped on and waited.

And waited.

Nothing happened.

After a short pause, Mark came on the two-way radio and told me the bad news: in the extreme heat the little electric winch we had counted on to haul me out of the pit was malfunctioning. He was going to have to think of something else. Happily for me, it there's anything Mark doesn't know about ropes and pulleys, it's probably not worth knowing. He quickly organised a hauler system and pressed a couple of our gun-toting escorts into helping him heave on the rope. I waited at the bottom for only half an hour. It seemed like forever.

In fits and starts, they got me back to the level of the cable, and I was able to clip on to the tether again. Then it was a simple matter of being hauled back along the wire. I was very happy to arrive.

'How was it down there?' Rachael said. I was dimly aware of the camera.

'It's pretty gnarly,' I reported. It's hard to be articulate when you're half-way to desiccation, but I managed a strained smile.

The following day, when the air force helicopter returned to extract us, what we had achieved began to sink in. I've done some demanding things in my time, but this was one of the hardest. I've done some mad things in my time, but this was undoubtedly the maddest. But it was another ambition realised. Chalk it up, and move on to the next one.

Dream Machine – Cyclone-proof Truck, Australia

I've always loved trucks, the bigger the better. I drove them for a living for a brief happy time before falling into my present line of work. Even after the job finished I kept up my subscription to *Trucking* magazine. So I couldn't believe my luck when Pioneer Productions agreed to bankroll the construction of a cyclone-chasing truck for an episode of *Dangerman*.

The camera – pointed first by Joe Morgan, and then by Rachael Wilson – followed me as I built my dream machine. First, I drooled my way through a second-hand truck yard in Perth, Australia. I found a couple of likely candidates, including one which was the kind of Oshkosh monster used for carting nuclear missiles about the countryside. Its last Australian owner had used it in gold-mining operations. But nothing I saw quite measured up to the MAN truck I'd seen on the internet, and which had drawn me to Perth in the first place. It would need a bit of modification to suit my purposes, but as soon as I saw it I knew it was the truck for me.

My cyclone-chasing truck, complete with huge steel mesh shutters and a vast array of satellite equipment.

Now that we had our 'tractor' (the section of an articulated truck which contains the engine), the next step was finding something on which to base the bodywork. It had to be strong enough to withstand whatever a Category

5 storm could hurl at it, and I knew enough about Cat-5 storms to know they could hurl just about anything that wasn't permanently fastened to the ground, and even some things that were. I settled upon the plate-steel cabin which had once graced the rear of a prison transfer vehicle.

The first garage I lined up to perform the overhaul and modifications failed to deliver, and for a while it looked as though nothing was ever going to happen, let alone in time to chase a cyclone that season or make the show's deadline. Just in time, my friend Daniel Lewis put me in touch with a rough-looking character named Flighty. Discovery Channel has a programme called *Monster Garage*, in which a team of mechanics morph ordinary cars into mega-machines. Well, this was *Dangerman's* answer to *Monster Garage*. With Flighty in control, the truck's engine and running gear were completely overhauled, the cabin fitted out with basic accommodation, and my storm-chasing communication gizmos installed. Huge steel mesh shutters were added to protect any vulnerable areas, and the whole thing was sprayed army green. The finishing touch was a slightly crazed array of orange flashing lights, and a definitely deranged locomotive horn that I imported from the United States.

The rest of the episode followed Rachael and I driving the truck north to its intended home, Broome, the cyclone capital of Australia. The journey wasn't without incident. First, I got the super-truck stuck while testing its four-wheel-drive in soft sand off-road. Then we broke down with a fuel problem hundreds of miles from anywhere. The Great Sandy Desert turned out to be not a great place to be marooned. It was indescribably hot, and while we waited for the mobile mechanic to make the day's drive out, our only water was what we carried. As if to illustrate how inhospitable the place was, a sorry-looking dingo staggered out of the desert and collapsed under the truck. We tried to give it water to drink, but it died overnight. When the local fauna, which is adapted to the conditions, start keeling over you know you're not in a pleasure spot.

The mechanic got us mobile again and we made it to Broome. But nothing in the way of cyclones came along that season – at least not in Australia. Shortly after we'd given up and flown home, Rachael and I diverted to Cyclone Heta in Samoa, where we shoot many dramatic storm scenes. My only regret was they didn't feature my new truck. Indeed, at the time of writing the truck has never chased a cyclone.

Its time will come. I can't wait.

GEOFF MACKLEY.COM

When the World Wide Web first became a practical and widespread tool in the 1990s, there was plenty of noise about its potential to revolutionise economics. People invested heavily in strange little companies floated to exploit this supposedly unlimited new potential. Millions were made overnight solely on the strength of the internet's promise – and just as quickly lost. The much-vaunted dot.com revolution has, by and large, turned out to be just another bubble.

But it has also had its success stories, and my career is one. Prior to the internet, I could not have done the things I now do every day, and if the internet for some reason ceased to exist I could not continue doing them.

The internet's chief value is the instant access it gives to real-time information on things happening in all parts of the world. How else could I find out about Atlantic storms from my home in Auckland? Even if I learned about a storm spinning up in the Azores High, how could I monitor its progress in order to decide whether to spend money chasing it? The sites that provide free, instant, up-to-date meteorological information are the lifeblood of my business.

The same goes for volcanoes and tornadoes. The entire twister-chasing scene is based on the operators' ability to access a range of different data about the rapid formation and evolution of storm cells in their immediate area. Without portable computers connected by the internet to weather radar sites, we would be no better informed about the likely formation of a funnel cloud than would someone listening to their radio or looking out their window.

Just as importantly, my own website, *www.geoffmackley.com*, provides a worldwide, 24-hour, 7-day-a-week advertisement for my work. I have thousands of clients whom I've never met but who regularly buy photographs and film footage from me. And whether I'm in the middle of the Sahara Desert or sitting in my living room at home, I can respond to requests and dispatch photos virtually instantaneously to anywhere around the world.

People often say I'm a lucky guy, and usually they're thinking of the kind of luck that keeps you from getting killed. But they're right for a different reason. The internet came along at precisely the moment I found myself in a position to consider travelling the world, getting amongst extreme natural phenomena. It came along with an inbuilt niche for what I do, just waiting for me to fill it.

Now that was lucky.

www.bom.gov.au/weather/wa/cyclone/index.shtml
Perth Tropical Cyclone Warning Centre *Tropical cyclone warnings*, forecasts, tracking and threat maps for northern Australia and Central Indian Ocean.

www.geonet.org.nz
The New Zealand GeoNet Project *Real-time monitoring* and data collection for rapid response and research into earthquake, volcano, landslide and tsunami hazards.

www.jma.go.jp/jma/indexe.html
Japan Meteorological Agency *Weather observations* and warnings. Earthquake, volcano and tsunami information and warnings.

www.maybagyo.com
David Michael V. Padua, Philippines *South-east and East Asia* tropical cyclone warnings, tracking and storm history.

www.metoffice.gov.uk/weather/tropicalcyclone/warnings.html
United Kingdom Meteorological Service *Tropical cyclone warnings* in the Indian and Pacific Oceans, with links to other warning and tracking websites.

www.metservice.co.nz
New Zealand Meteorological Service *Provides weather information* from New Zealand, around the world.

www.nhc.noaa.gov
US National Weather Service National Hurricane Center *Forecasts and information* about hurricanes across the United States.

www.npmoc.navy.mil/jtwc.html
Naval Pacific Meteorology and Oceanography Center Joint Typhoon Warning Center *Issues tropical cyclone* warnings for the Pacific and Indian Oceans.

www.nrlmry.navy.mil/tc_pages/tc_home.html
NRL Monterey Marine Meteorology Division Tropical Cyclone Page *Atlantic, Pacific and Indian Oceans* storm tracking and information.

www.osei.noaa.gov
National Environmental Satellite, Data, and Information Service *High-resolution, detailed* imagery of significant environmental events, including tropical cyclones, hurricanes and typhoons, volcanic eruptions and fires.

www.sphere.net.au/strikestarau/
Strikestar *Networked lightning* locator system.

www.sveurop.org/gb/news/news.htm
Weekly updates on worldwide volcanic activity.

www.volcano.si.edu/reports/index.cfm
Smithsonian National Museum of Natural History *Weekly reports and history* of worldwide volcanic activity, issued for individual volcanoes.

www.wunderground.com/auto/HurricaneSite/tropical/
Tropical weather forecasts, tracking and history in the Atlantic and Pacific Oceans.

Glossary

ashfall Descent of fine, dust-like volcanic material ejected during a volcanic eruption.

caldera Large crater of a volcano in which individual vents are located.

CAPE Convective available potential energy: the energy present within a **storm cell** which creates a **vortex** and rotation, and therefore one of the measurable characteristics of a storm cell that gives early warning of tornado formation.

carabiner Specialised clip used by climbers and mountaineers.

Coriolis effect Named for G. de Coriolis, the scientist who first described it, the general effect observable in the motion of objects relative to a rotating frame of reference, but more particularly used to describe the rotation of weather systems – clockwise in the northern hemisphere and anti-clockwise in the southern hemisphere.

cumulonimbus Clouds, similar in form to **cumulus** cloud, but towering to great altitude. A typical formation in thunderstorms.

cumulus Cloud-type consisting of rounded masses heaped above a low-altitude base.

eyewall Dense, cylindrical mass of cloud surrounding the clear 'eye' of a hurricane or cyclone, often extending to great altitude. Windspeeds are greatest in the eyewall of a hurricane, even though in the eye there may be total calm.

fumarole Vent, usually associated with a volcano, through which superheated gases emerge.

hectopascal (hPa) Metric unit used in measuring atmospheric pressure (one hPa corresponds to 100 pascals, or one millibar, its Imperial counterpart).

hornito Small volcanic vent surrounded by a heap of volcanic material.

lahar Volcanic landslide, usually consisting of superheated mud, water and assorted debris.

lava dome Mass of magma extruded from a subterranean volcanic vent and crusted over with solidified material. *Also known as* cryptodome.

magma Molten rock before it emerges from a volcanic vent.

Magnitude scale (M_w) Logarithmic scale used to describe the intensity of seismic activity. A logarithmic scale is one where each term is ten times greater than the preceding term. For example, a magnitude 5 earthquake is ten times stronger than a magnitude 4 earthquake.

mesoscale convective complex Technical term for a weather system whose distinguishing feature is an area of convection — rapidly rising, warm air.

mesocyclonic cells Areas of convection within a thunderstorm which give rise to rotation – in other words, the precursor to tornado formation.

nevée Catchment of a glacier, consisting of snow before it has been compressed into ice.

nuée ardente Literally, 'glowing cloud', a mass of superheated gas and particles emitted in a certain type of volcanic eruption, invisible in a pall of ash and smoke during the day, but glowing intensely at night (hence its name).

orographic cloud Striking, ribbon formation of cloud in the lee of a mountain or range of mountains.

peléan eruptions Sudden, violent volcanic eruptions with little seismic activity in between. Named for Mount Pelée, which is characterised by this type of activity.

phreatic eruption Volcanic explosion caused by the heating and resultant expansion of water underground.

plate tectonics Geological theory whereby the Earth's crust is formed of plates 'floating' on the semi-liquid mantle beneath. The plates are thought to be in motion, and it is their collision that gives rise to vulcanism and seismic activity (earthquakes).

pumice Light, porous rock formed when silicon-rich volcanic lava is cooled rapidly, as in contact with water.

pyroclastic flow Mass of superheated debris ejected from a volcano during eruption and moving down its sides, often at very high speed.

Richter scale Now obsolete scale for measuring the intensity of earthquakes, named for its inventor, Charles Richter. It has been replaced by the **Magnitude scale**.

scoria Porous rock formed when lava cools slowly.

shield volcano Type of volcano formed around a vent or vents from which volcanic material is generally emitted sluggishly.

sink-hole Vertical cavern, commonly found in limestone landscapes, where water has eroded a pocket of softer material over time to form a cavity in the harder material surrounding it.

storm cell Area of high energy within a thunderstorm, usually characterised by a pair of convective columns of air, one consisting of warm, rising air, the other of cooled, falling air. The electrical charge developed by storm cells gives rise to thunder and lightning.

stratovolcano Type of volcanic cone created by successive eruptions of varying types.

Strombolian activity Volcanic activity characterised by continuous, low-level eruptions of glowing, molten material.

supercells Very high-energy **storm cell**, sometimes giving rise to tornadoes.

tectonic plates Individual segments into which the Earth's crust is divided, according to the theory of **plate tectonics**.

tectonic subduction Collision of **tectonic plates** at a boundary where the edge of one plate is forced underneath the edge of the other.

tephra Gravel-like debris ejected from an erupting volcano.

The Roaring Forties Area of the Earth's surface lying between latitudes 40° and 50° S, notorious for its tumultuous weather.

thunderheads, High, stacked cumulonimbus clouds associated with electrical storms. *Also known as* thunder clouds.

troposphere Lowest layer in the Earth's atmosphere, in which all weather phenomena occur.

vortex Spiralling of liquid or gas around an area of low or nil pressure caused by the **Coriolis effect.**

Index

CHEVROLET